Asset Analytics

Performance and Safety Management

The main aim of this book series is to provide a floor for researchers, industries, asset managers, government policy makers and infrastructure operators to cooperate and collaborate among themselves to improve the performance and safety of the assets with maximum return on assets and improved utilization for the benefit of society and the environment.

Assets can be defined as any resource that will create value to the business. Assets include physical (railway, road, buildings, industrial etc.), human, and intangible assets (software, data etc.). The scope of the book series will be but not limited to:

- Optimization, modelling and analysis of assets
- Application of RAMS to the system of systems
- Interdisciplinary and multidisciplinary research to deal with sustainability issues
- Application of advanced analytics for improvement of systems
- Application of computational intelligence, IT and software systems for decisions
- Interdisciplinary approach to performance management
- Integrated approach to system efficiency and effectiveness
- Life cycle management of the assets
- Integrated risk, hazard, vulnerability analysis and assurance management
- Adaptability of the systems to the usage and environment
- Integration of data-information-knowledge for decision support
- Production rate enhancement with best practices
- Optimization of renewable and non-renewable energy resources

More information about this series at http://www.springer.com/series/15776

Hassan Qudrat-Ullah · Pramela Nair Panthallor

Operational Sustainability in the Mining Industry

The Case of Large-Scale Open-Pit Mining (LSOPM) Operations

Springer

Hassan Qudrat-Ullah
College of Industrial Management
King Fahd University of Petroleum and Minerals
Dhahram, Saudi Arabia

Pramela Nair Panthallor
Academic Doctoral Consultant-Freelance
North York, ON, Canada

ISSN 2522-5162 ISSN 2522-5170 (electronic)
Asset Analytics
ISBN 978-981-15-9026-9 ISBN 978-981-15-9027-6 (eBook)
https://doi.org/10.1007/978-981-15-9027-6

This Springer imprint is published by the registered company Springer Nature Singapore Pte Ltd.
The registered company address is: 152 Beach Road, #21-01/04 Gateway East, Singapore 189721, Singapore

To my husband, Suresh Kalathil, for always encouraging, motivating, and showing immense interest in understanding, listening, and sharing his thoughts about my work. Also, to my daughter, Avantika Nair, who has been a strong support throughout my research, inspiring me to do the best in every project I undertake

—Pramela Nair Panthallor

To Anam Qudrat, my daughter, who continues to pleasantly amaze me through her intellectual curiosity and dialogues (often heated)—the lifeline for the well-being and prosperity of our large family and our aspirations and pursuits.

—Hassan Qudrat-Ullah

Preface

Minerals and metals form a vital ingredient in almost every product used in the human value chain. Their demand is ever increasing. Many national economies depend heavily on the mining industry for their economic growth. High costs and increasing technology need pose serious challenges for sustainable mining operations. The stakes in the mining industry are therefore high and consumption of mining resources requires the resources to be used efficiently.

The objectives of this book are threefold: (i) to better understand the relevance of efficiency and effectiveness in mining, (ii) the requirements to operate it sustainably for optimal use of resources, and (iii) to understand the significance of organizational efficiencies, information technology, and decision-making in the mining operations.

These objectives were accomplished through a rigorous research project involving intensive consultation with the mining industry professionals including executives in the top-level management belonging to 12 different nationalities from 14 organizations around 10 countries and experience working in different types of metals. By applying a mixed-method research approach, our analysis provides unique insights into the mining industry in general and in LSOPM operations.

Chapter 1 introduces the world of mining, its origin, its relevance to modern industry, and personal use along with its major contributions to various national economies, describes the mining phases and processes. Chapter 2 deals with a detailed understanding of information technologies and systems present in the industries and emphasize information systems within the LSOPM industry. Chapter 3 presents the concepts of organizational efficiencies in terms of costs, productivity, equipment, manpower, time, and performance in general and in terms of LSOPM operations. This chapter emphasizes the mining equipment and its concepts of availability, utilization, and maintenance. Chapter 4 provides a detailed understanding of types of decisions, individual and group decision-making, decision-making processes and models, strategic decision-making, effective managerial decision-making, and tools and techniques used for group decision-making. Apart from that, the chapter shows the advantages of better decision-making between the interrelationship of swift decision-making and productivity in LSOPM operations. Finally, Chapter 5 concludes with new models that can be used for the operational sustainability of mining operations and further research area considerations.

We believe this book will benefit students pursuing undergraduate and graduate studies and the professionals in the mining industry to understand the primary relevance of various efficiency factors in an organization. For instance, this book can serve as a supplementary book for the courses Mine Management in the Mining Engineering discipline, Operations Management/Production Management in the Business Administration discipline, and in general for all courses, which include organizational efficiency as part of the syllabus.

Toronto, Canada

Pramela Nair Panthallor
Hassan Qudrat-Ullah

Acknowledgments

We are thankful to many individuals who have contributed to the completion of this book project. Pramela Nair Panthallor would like to appreciate her thesis supervisors from Monarch Business School for providing prompt and constructive feedback and inspiring and motivating her throughout her doctoral thesis that leads to the completion of this book project. She would like to extend special thanks to her dear parents and sister who believed in self-dependence and considered education to be especially important in pioneering changes in the thought processes of individuals and bringing in a major advancement in the society.

Thanks to people from Springer, especially Nupoor Singh, who provided exceptional help, support, and encouragement during the review and contracting processes of this book.

Finally, Hassan Qudrat-Ullah would like to acknowledge the financial support provided by the Deanship of Research (DSR) at KFUPM for funding this through project #: BW193-MGTMKT-97.

Contents

About the Authors

Hassan Qudrat-Ullah is a professor of decision sciences, at the School of Administrative Studies, York University, Canada. Dr. Hassan has over 20 years of teaching, research, industry, and consulting experience in the USA, Canada, Singapore, Norway, UK, Korea, China, Saudi Arabia, and Pakistan. His research interests include dynamic decision-making, interactive learning environments, system dynamics, and energy policies. His research has appeared in *Energy Policy*, *Energy, Simulations & Gaming*, *International Journal of Management and Decision Making*, *Journal of Decision Systems*, *Telecommunication Systems*, *International Journal of Technology Management*, *Decision Support Systems*, and *Computers & Education*. Hassan has also published 11 books and over 100 refereed articles in journals and conference proceedings.

Pramela Nair Panthallor completed her Doctoral study in Business Research from Monarch Business School, Switzerland, her MBA from Sikkim Manipal University, India, and her B.E. in Computer Science from Orissa Engineering College, India. She was awarded a distinction for her doctoral thesis work and was chosen through peer and student evaluations for an award of Most Excellent Lecturer (2016–17) for outstanding teaching efforts. She has more than 8+ years of experience in teaching undergraduate courses in International Business, Software Engineering, and Computer Science & Engineering. She has lived in seven different countries and has experience working in five of them.

Chapter 1
Introduction: Managing Large-Scale Mining Operations

1.1 Introduction

The global mining industry is a critical component of a multitude of value chains extending across continents (ICMM, 2018). Metals and minerals extracted by mining serve as an enabler for various fields like farming, healthcare, communications, water, and energy supply, transport, space technology, and the construction of cities. Mining as an industry globally accounts for approximately USD 735 billion of extracted mineral per year and is one of the world's largest industries (Ericsson & Larsson, 2013). As the basic ingredient to produce products efficiently, it is a necessity that mining companies optimize costs and efficiencies throughout the production chain (Van Niekerk, 2013).

Like any other industry, the mining industry also goes through the business cycle with the recession and boom periods. When there are a recession and the commodity prices in the mining industry, i.e., the metal prices are low, generally, all mining companies try to see what the waste in their processes is and to eliminate or minimize the same to maximize the cost optimization. However, this exercise should continue regardless of whether there is a recession or not. For example, the drop in the market price for gold in Oz from the highest price of USD 2,076.25 in 2011 to the lowest price of USD 1,153.22 in 2015 is a huge difference for 4 years (Macrotrends, 2020). These kinds of price differences require companies to optimize processes to make sure that both the mining, process plant, and every other department work together in achieving efficiency in terms of equipment, manpower resources, time, cost, and sustain productivity.

One of the developments that changed the outlook for the availability of metals is the rapid advances in technology including the production of minerals from lower grade ore (Mikesell & Whitney, 2016). Information technology is a critical driver that mining companies use to optimize costs and improve efficiencies for better decision-making (ICMM, 2015). Every technology is different and there exists a difficulty in determining the value-added to the business by any newly acquired technology or

H. Qudrat-Ullah et al., *Operational Sustainability in the Mining Industry*,
Asset Analytics, https://doi.org/10.1007/978-981-15-9027-6_1

process. The demand for metals is increasing significantly, the pressure to produce huge volumes, and the mining industry is one of the most energy-intensive industries that resulted in huge energy demand. Excluding energy consumption of processes in metallurgy, transportation, and other activities related to mining, according to the International Energy Agency, around 8–10% of the world's total energy consumption is related to mineral extraction as per the requirement of the society for minerals (Calvo et al. 2016). This chapter is further divided into three sections:

- Mining Origins
- Relevance of Mining
- Mining Definitions and Processes

1.2 Mining Origin

One of the main aspects of the advancement of human history is the development of tools for ages with the discovery and understanding of various metals and minerals from the stone age to the industrial revolution. In history, mining is generally related to people rushing to different geographical regions, hoping to get rich quickly by discovering precious metals (InEight Inc., 2013). However, the history of mining is also associated with the evolution of mankind and the advancement of usage of weapons throughout history for metals like copper, bronze, iron, and steel and do not mention the historic details on many other metals such as aluminum, nickel, and manganese (Coulson, 2012; Hartman & Mutmansky, 2002; Snedeker, 1990).

Mineral resources are finite in nature and are formed by natural earth processes (Sykes et al., 2016). Mining of minerals dates about 450,000 years (Hartman & Mutmansky, 2002) with considerably basic processes done on a small scale in most parts of the world until the fifteenth century (Burt, 1991). Machines and new techniques arrived much later, after which the concept of large-scale mining was considered feasible as an investment. Changes in mining techniques occurred in a random and unplanned manner due to the impact of innovations in technology, in terms of electricity, blasting using explosives, and the introduction of machines and information technology (Burt, 1991). In the past, the estimation and worth of a mining investment depended largely on the way it was interpreted by the mining engineers and geologists (Nicholas, 1910). This had a huge impact on the return on investments in mining projects.

The history of mining dates to the stone age, also known as the Palaeolithic period, when the soil was extracted by humans to make tools and weapons (Hartman & Mutmansky, 2002). But one of the first industrial experiences for mankind was mining and the use of first tools and weapons used by the prehistoric men made of stone (Coulson, 2012). Copper and gold were the first metals to be used and were used by people residing in the valleys of Tigris, Euphrates, and Nile (Graham & Evans, 2007).

Breaking the ore and loosening it from the surrounding rocks was one of the first challenges for early miners and the crude tools used by them which were made of

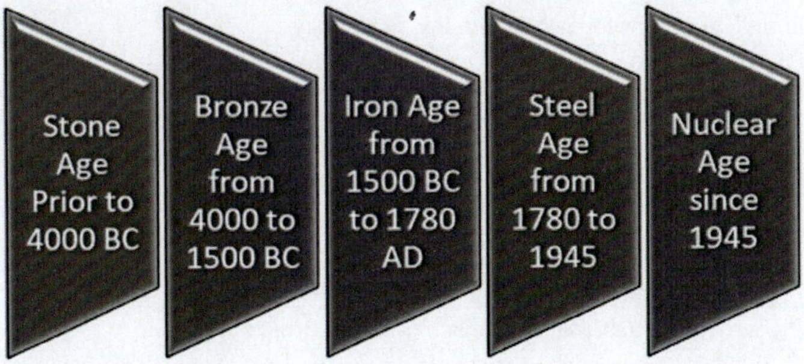

Fig. 1.1 Different periods in history according to the mineral used for weapons

bone, wood, and stone were no match for the hard rocks (Hartman & Mutmansky, 2002). In the Stone and Bronze Ages, tools made by stone implements were used for digging beside the rocks and breaking it. Using bronze did not help because it was too valuable and too soft to be a substitute for stone. Copper was used for making tools and weapons in Mesopotamia and Egypt, and later this knowledge was embraced across Europe, during the 3000 s BCE (Graham & Evans, 2007). Minerals were discovered according to the evolution of mankind in the usage of weapons (Snedeker, 1990) as shown in Fig. 1.1.

Copper replaced bronze in 1500 BC, which improved the quality of weapons and tools and when the strength of iron was realized, it substituted for bronze after around 100 years from then. Initially, metals were used in their more natural forms, but during the Bronze and Iron Ages, the smelting of metals was discovered by humans (Graham & Evans, 2007). Humans learned to convert ores into pure metals or alloys, improving the ability to use these metals (Hartman & Mutmansky, 2002). However, in terms of efficiency, it was during the Iron Age that mining took a step forward by introducing several important metallic compounds that could be used to make weapons and other durable tools—namely copper, brass, bronze, and iron—which played a huge role in the advancement of civilization (Coulson, 2012).

The timing of the history of metals was not the same everywhere around the world (Sykes et al., 2016). For example, while the Americas only had copper, Africa's Iron Age had already started, the reason of which has resulted in the availability of appropriate mineral resources and technological understanding. Economies cannot progress rapidly without access to new findings of the high quality of metal deposits, better techniques of metal extraction, and renewed uses of metals. The development of mining technology (Hartman & Mutmansky, 2002) is given in Table 1.1 in chronological order.

Minerals and metals are used in manufacturing, services, and infrastructure in modern society, which include food and water supply, shelter, energy supply, sewage treatment, transportation, construction, education, health, communication, entertainment, tourism, and a huge range of related consumer goods and services (ICMM, 2012b). Until the seventeenth century, rock breaking was done manually, while years

Table 1.1 Chronological development of mining technology

Date	Event
450,000 BC	First mining (at the surface), by Palaeolithic man for stone implements
40,000	Surface mining progresses underground, in Swaziland, Africa
30,000	Fired clay plots used in Czechoslovakia
18,000	Possible use of gold and copper in the native form
5000	Fire setting, used by Egyptians to break rock
4000	Early use of fabricated metals; the start of Bronze Age
3400	First recorded mining, of turquoise by Egyptians in Sinai
3000	Probable first smelting, of copper with coal by Chinese; first use of iron implements by Egyptians
2000	Earliest known gold artifacts in New World, in Peru
1000	The steel used by Greeks
100	Thriving Roman mining industries
1185 AD	Edict by the bishop of Trent gives rights to miners
1524	First recorded mining in New World, by Spaniards in Cuba
1556	First mining technical work, *De Re Metallica*, published in Germany by Georgius Agricola
1585	Discovery of iron ore in North America, North Carolina
The 1600 s	Mining commences in the eastern USA (iron, coal, lead, gold)
1627	Explosives first used in European mines, in Hungary
1646	First blast furnace installed in North America, in Massachusetts
1716	The first school of mines established, at Joachimstal, Czechoslovakia
1780	Beginning of industrial revolution; pumps first modern machines used in mines
1855	Bessemer steel process first used, in England
1867	Dynamite invented by Nobel applied to mine
1903	The era of mechanization and mass production opens in US mining with the development of the first low-grade copper porphyry, in Utah; although the first modern mine was an open pit, subsequent operations were underground as well

later gunpowder was used for blasting, which is considered as the biggest innovation in mining since it resulted in a great increase in productivity (Burt, 1991). Even though manual loading and hauling were done hundreds of years ago, the machines that replaced it still required manual reporting systems with information on availability, maintenance, and position information.

Mining was considered tedious and dangerous before the eighteenth century until the industrial revolution, which resulted in steam-powered pumps, drills, and lifts; but currently, technology makes mining a safe and efficient option as an industry (InEight Inc., 2013). However, the advancement in breaking rocks came with a revolutionary technique called fire setting, in which the rocks were first heated so that they expand and then immersed in cold water so that they contract and break. And later, the invention of dynamite by Alfred Nobel in 1867 made rock breakage an easy phase of mining (Hartman & Mutmansky, 2002).

1.2.1 Relevance of Mining

Mining materials are categorized by the US Bureau of Mines and other mining agencies into three main groups: minerals, metals, and fuels (Deneen & Gross, 2009). Of the three groups, metal mining is most difficult, physically, and metals of different types are usually found in the same ore. The basic characteristics of minerals, specifically, metals due to its toughness, durability, conductivity for heat and electricity, beauty, and reasonable cost, making them an important commodity (ICMM, 2018). The symbols of prominent metals and minerals and the relevance of various metals (ICMM, n.d.) and minerals in day-to-day life (ICMM, n.d.) are given in Tables 1.6 and 1.7 respectively, in the appendix section at the end of the chapter.

The scale and cost are the reasons why metals have a high advantage over other materials that can replace and substitute it (Deneen & Gross, 2009). Metal mining consists of the extraction of minerals like copper, nickel, zinc, and other basic metals including the precious metals, such as gold, silver, and platinum. The demand for metals can be expected to be extremely high due to its continuous use in day-to-day life. Demand for the metal is related to factors like per capita income, for example, demand for metal increases very quickly when per capita income of any country is about USD$5,000 to USD$10,000 per year which is why, due to high population rate, countries like India and China show dramatic effects in demand when they go through this development phase (ICMM, 2012a). The sales in million USD for different metals (Tilton & Guzman, 2016) are shown in Table 1.2.

The minerals and metals that find themselves in demand for the present and future are coal, iron ore, copper, bauxite, phosphate, potash, nickel, zinc, and lead. For example, the demand for iron by 2030 is around 3,500 million tonnes per year, and for copper and nickel around 28 million tonnes and 3.8 million tonnes, respectively (Wenco, 2015). Premier metals like iron ore, copper, gold, and nickel are believed to remain particularly important investments for mining companies accounting for 84% of the total future projects (Wenco, 2015). The market growth percentage of different metals from 2004 to 2013 (Sykes et al., 2016) is given in Table 1.3.

The metals are cumulatively valued at USD 606 billion globally, which constitutes to 71% of the total value of all nonfuel metal and mineral production in 2011 (ICMM, 2012b). The demand for some of the metals produced globally for the year 2015 (Calvo et al., 2016) is given in Table 1.4 in the appendix section. Minerals and metals have been of vital importance due to a much wider and broader range of uses from capacity in metals for recycling to its ability in providing employment for skilled and unskilled labor for mining activity (ICMM, 2014). This can also be seen in the way metals could be used for infrastructure development and a list of applications of metals like aluminum boats, copper in wind turbines, and chromium–cobalt alloys used in the artificial hip (ICMM, 2014).

Table 1.2 Statistics on sales of various metals globally for the year 2014

Type of metals	Classification of metals	Metals	2014 sales (million $)
Ferrous		Iron Ore	325,220
Nonferrous metals	Base	Aluminum	120,238
		Copper	129,862
		Tin	6,395
		Nickel	40,471
		Lead	10,352
		Zinc	28,881
	Light	Beryllium	121
		Lithium	1,265
		Titanium	4,982
	Platinum Group Metals (PGM)	Iridium	582
		Palladium	5,070
	Precious	Gold	116,778
		Platinum	7,454
		Silver	15,969
	Refractory	Cobalt	3,457
		Molybdenum	7,155
		Niobium	3,812
		Rhenium	146
		Tantalum	355
		Tungsten	3,637
		Zirconium	1,494
	Nonclassified	Arsenic	30
		Bismuth	213
		Cadmium	43
		Chromium	6,467
		Indium	582
		Manganese	36
		Mercury	100
		Silicon	26,530
		Strontium	16
		Vanadium	1,780

Table 1.3 Ranking of market growth for metals from 2004 to 2013

Rank	Metals	Market growth (%)	Rank	Metals	Market growth (%)
1	Lithium	388	21	Manganese	105
2	Germanium	355	22	Zinc	105
3	Gallium	327	23	Rare Earth Metals	93
4	Mercury	320	24	Tantalum	74
5	Tungsten	298	25	Niobium	72
6	Rutile	295	26	Magnesium	62
7	Silver	271	27	Indium	57
8	Gold	221	28	Cadmium	53
9	Beryllium	215	29	Ilmenite	51
10	Bismuth	214	30	Tin	50
11	Antimony	211	31	Aluminum	44
12	Lead	190	32	Vanadium	24
13	Rhenium	171	33	PGM's	19
14	Barite	169	34	Nickel	12
15	Uranium	156	35	Cobalt	−5
16	Copper	155	36	Molybdenum	−18
17	Thorium	146	37	Arsenic	−30
18	Chromium	144	38	Borate	−57
19	Silicon	118	39	Strontium	−57
20	Tellurium	116			

Table 1.4 World production and resource availability of some metals for the year 2015

#	Metals	Production (in tonnes)	Available resources (in tonnes)	Identified resources (in tonnes)
1	Copper	18.7 million	720 million	2.1 billion
2	Lead	4.7 million	89 million	2 billion
3	Zinc	13.4 million	200 million	1.9billion
4	Gold	3000	56,000	135,000

The benefits of mining are not limited to individuals or specific industries only; it also contributes in a major way toward many national economies. The mining industry includes many stakeholders like the government, investors, contractors and suppliers, service providers, communities affected by mining, other organizations, labor unions along with research and academic institutions (ICMM, 2012b). Studies show that each job created by mining leads to the creation of three to five other jobs outside the mining sector (ICMM, 2014). In 2013, it was estimated that in the USA alone around 270,000 people work in mining and the mining industry created three million related jobs (InEight Inc., 2013).

ICMM releases a Mining Contribution Index (MCI) that signifies the mining sector's contribution to national economies. The ranks provided through MCI show the relative importance of mining to the economic life of a country. The MCI shows that low- and middle-income countries significantly depend economically on mining. Suriname, Democratic Republic of Congo, Guinea, Burkina Faso, and the Kyrgyz Republic are in the top five ranks according to the 2018 rankings (ICMM, n.d.). The MCI rank of the top 25 countries for the year 2018 is given in Table 1.8 in the appendix section at the end of the chapter.

Countries that were dependent on mining and mineral resources over the entire period of 1995–2015 are Botswana, Central African Republic, Niger, Mauritania, Ghana, Guinea, Togo, Zambia, Democratic Republic of Congo, Namibia, South Africa, Mongolia, Bolivia, Chile, Guyana, Jamaica, Peru, Suriname, Armenia, and Georgia (ICMM, n.d.). In terms of the monetary value of metals and minerals production, the ranking shows that the upper and middle-income countries like China, Australia, Russia, USA, and India are at the top five rankings, respectively (ICMM, n.d.). The production value of metallic mineral (including coal) for the year 2016 of the top 20 countries is given in Table 1.9 in the appendix section at the end of the chapter. Table 1.5 and Fig. 1.2 show the contribution of export in the mining sector toward the GDP in 20 countries for the year 2012 (ICMM, 2012b).

Table 1.5 Export contributions of minerals and metals (coal excluded)

Rank	Country	GDP/capita (PPP at current prices, 2012 US$)
1	Botswana	14,707
2	Democratic Republic of Congo	697
3	Suriname	15,440
4	Mongolia	8,442
5	Zambia	3,043
6	French Polynesia	n/a
7	Mauritania	2,878
8	Chile	21,045
9	Eritrea	1,200
10	Guinea	1,237
11	Peru	11,103
12	Tajikistan	2,361
13	Guyana	6,159
14	Namibia	9,316
15	Papua New Guinea	2,424
16	Sierra Leone	1,610
17	Burkina Faso	1,555
18	Sudan	3,607
19	Montenegro	13,528
20	Armenia	7,418

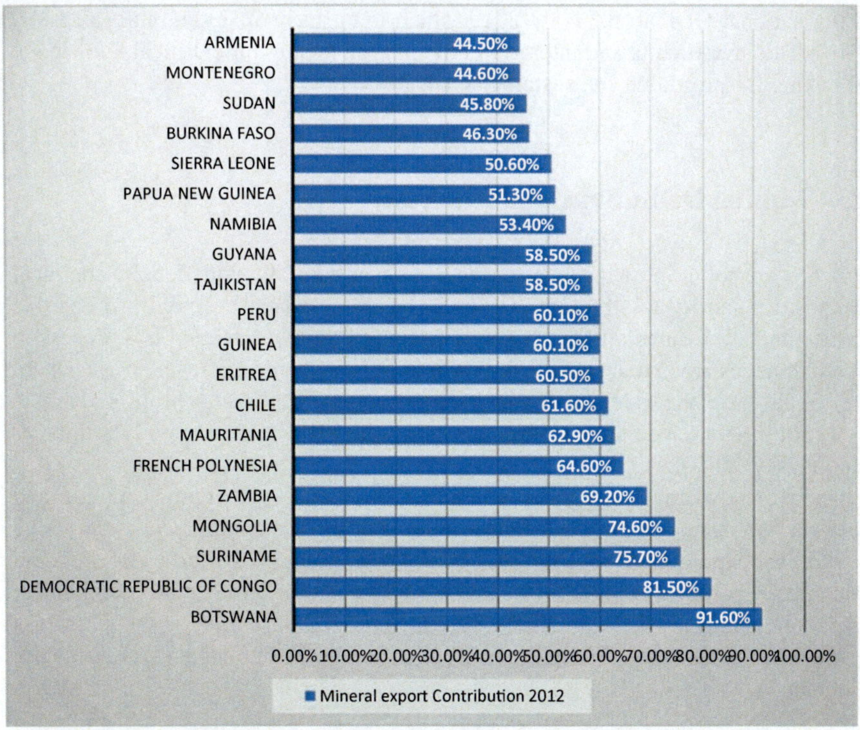

Fig. 1.2 Mining export contribution to national economies (year 2012)

Large investments in mining lead to the requirement of developing business models to justify if a decision of investing in information technologies leads to cost–leadership strategies. Beck Nader, executive director of Australian Software and Consulting Company, Micromine, South America, explains that in the case of small-scale miners there is an educational issue involved due to the specialized technologies that are new and expensive (Minas Gerais, 2012). Implementation of high-end information technologies is generally accomplished, specifically, in large-scale mining companies that expect to benefit from the use of enormous data available from different streams of management- and operator-level input.

Mining is an industry that is capital intensive; in fact, various countries and organizations have significant investments and interest in mining. Andres Poch, the President of the Association of Consulting Engineers of Chile, stated that over the next few years around US$120 billion will be invested in mining (E&MJ, 2012b). The Russian mining company, Alrosa, has a major investment project at the Zarya open pit for which the initial investment is planned at about US$150 million (E&MJ, 2012b). A plan to conduct a survey aerially of 800,000 km^2 of land by the ministry of mines in India shows immense geological potential in India (E&MJ, 2016). The survey is conducted to boost the contribution of India's domestic mining industry to the national GDP to 3.3% by 2019 which is up from 2.3% currently (E&MJ, 2016). The investments and interests in the mining area require critical knowledge of making it a profitable and sustainable business.

1.3 Mining Definitions and Processes

Minerals are solid, inorganic substances that occur naturally with definite chemical composition and crystal structure (ICMM, 2018). Minerals can be found on the earth's surface and are formed without the intervention of humans. *Mining* is a process in which minerals are extracted from the earth containing a mix of ore and waste where ore should have the value with which its extraction results in a profit and waste is everything that accompanies the ore during the extraction process (Hartman & Mutmansky, 2002).

There are two main categories of mining, which are underground mining and open-pit mining, also known as opencast mining or surface mining. *Open-pit mining* is where the extraction of ore from rock bodies takes place on the surface of the Earth using heavy equipment like drills, shovels, and trucks for efficient and economical operations (Hartman & Mutmansky, 2002). *Underground mining* is the extraction of ore below the surface through subsurface openings for human entry (Hartman & Mutmansky, 2002).

1.3.1 Open-Pit and Underground Mining

Underground mining for gold by Egyptians was done 4,000 years ago; it is speculated that the Persians, Greeks, and Romans learned gold mining from the Egyptians (Graham & Evans, 2007). During the beginning of the twentieth century, most of the mines in the developed countries were underground mines but later as mining developed gradually in emerging economies, many mines in the world turned to be open-pit mines. Underground mining is complex and requires shaft sinking for the conveyance of people, material, and equipment. Open-pit mining, however, is comparatively not as expensive and is safer than underground mining.

Open-pit mining accounts for around 90% of the mineral resources mined in the USA. The concept of large-scale open-pit mining did not exist until the seventeenth century and it was generally performed by a small group of people (Net Industries, 2016). Global metal ore production is around 6,000 million tonnes per year and open-pit mining accounts for around 83% of this while underground methods account for the remaining 17% (Ericsson, 2012). In mining, the small companies and individuals are usually related to the exploration part of the industry while bigger companies in the world give attention to mining production and operations for higher profit generation

(Statista, Inc., 2016). Roughly, 52% of operations in metal mines globally are open-pit operations, 43% are underground operations, and the rest are tailing operations. Excluding the small-scale mines, it is estimated that in terms of "industrial-scale" operations, there are around 2,500 metal-producing mines in the world (Ericsson, 2012).

Copper ore found in 1903 in Bingham Canyon near Salt Lake City in the USA was one of the first mines operated as a large-scale open-pit mine to become profitable even though many believed it would not due to its low grade of 2% (Fiscor, 2016a). The Utah Copper Company was formed by Daniel C. Jackling for mining and processing the copper ore in Bingham Canyon. A huge open-pit mining operation was developed where shovels run by steam shovels loaded railcars to transport the ore to a large mill outside the canyon (Fiscor, 2016a).

In open-pit mining, to access the ore body, an opening is made to a large stretch of ground so that the ore is exposed to air. In the beginning, a small pit in the surface is developed which opens to a larger pit in a way enclosing the small pit—a process that continues until the final pit (Sevim & Lei, 1998). As the ore or waste is excavated and removed from the surface of the land, a deeper pit is formed in open-pit mining (Amankwah et al., 2014). However, the deep cone-shaped holes or pits in open-pit mines are first loosened by blasting. The extraction of ore is completed by sequentially digging the ground layers so that a cup-shaped pit is created, which has a staircase structure on the walls for ensuring the transportation of the materials removed from the pit (CATAPA, 2016).

Open-pit mines must be widened continuously as they go deeper for preventing the sides from collapsing. Furthermore, due to the size of open-pit mines, it often takes many years to excavate, which makes it an expensive process to reclaim the pits later by filling them with rocks (Net Industries, 2016). Figure 1.3 shows one of the open-pit copper mines of Rio Tinto in Chile (Mining Journal, 2016).

1.3.2 Life of Mines in LSOPM

A large open-pit mine that has a life of many years in terms of the LOM requires successful optimization techniques to handle the complex and enormous task of operating, planning, and management of the mine (Caccetta & Hill, 2003). The number of stages or phases in LOM differs among various studies, although they share many similarities. In general, the LOM can be categorized into five stages as shown in Fig. 1.4.

The first stage of prospecting deals with searching for ore for mining (Hartman & Mutmansky, 2002). The method of prospecting can be divided into two methods—the direct method and the indirect method. In the direct method, geologists conduct visual identification by means of fieldwork or aerial photography while in the indirect method, relatively hidden ore bodies are searched by specialists in the field of geophysics and geochemistry (Miguel, 1996). Traditional methods of prospecting

Fig. 1.3 Open-pit copper mines of Rio Tinto in Escondido, Chile

Fig. 1.4 Life of mine

have been replaced by micro quantitative analysis of water and soil samples, geobotany, gravitational, seismic, magnetic, electrical, electromagnetic, and radiometric measurements (Miguel, 1996).

The second stage is the exploration stage where the extent and value of ore examination/evaluation are defined (Hartman & Mutmansky, 2002). In the exploration stage, direct sampling of the area is done, usually by drilling, and then the samples are analyzed by geologists to assess the metallurgical properties. After analysis and interpretation, the ore body model is formed in terms of volume and orientation (Miguel, 1996).

The third stage is the development stage where the ore is exposed to production to commence (Hartman & Mutmansky, 2002). The development stage is the capital expenditure stage in which the mine is designed, the infrastructure is installed, and the plant, buildings, major equipment, roads, and processing materials are all incorporated into the mine design (Miguel, 1996). The fourth stage is the exploitation stage, which deals with the actual production of ore from the mines (Hartman & Mutmansky, 2002). In this stage, all resources are employed and many economic factors such as metal price, mineral demand, and the mining tax rate changes play a vital role in sometimes requiring changes in the original mine plan (Miguel, 1996).

The last stage of the life of mine is the reclamation stage where the mining site is restored (Hartman & Mutmansky, 2002). In the reclamation stage, the mining has concluded, and the mining area needs to be reclaimed into environmentally acceptable conditions. This stage is an on-going process and includes following mandatory government regulations and a considerable amount of capital deployment (Miguel, 1996).

1.3.2.1 Other Interpretations of Phases of LOM

Apart from the five phases detailed above, other commonly used interpretations for LOM are given below.

1. Seven-phase LOM (Kirsch et al, 2012): The seven phases for LOM are (a) exploration and planning, (b) construction, (c) operation, (d) decommission, (e) closure, (f) post-closure and (g) re-commission. Contradictorily, some studies mention eight phases in LOM: (a) exploration, (b) development, (c) active mining, (d) disposal of overburden and waste rock, (e) ore extraction, (f) beneficiation, (g) tailings disposal, and (h) site reclamation and closure.
2. Eight-phase LOM (ELAW, 2010): The eight phases of LOM can otherwise be explained in the following ways:

(a) During the exploration phase, the information in terms of survey results, field studies, results from drilling test boreholes, and other exploratory excavations related to locations and value of the ore is obtained.

(b) The development phase is continued if the exploration stage proves that there is a huge mineral ore deposit of minimum required grade. The development phase includes the construction of access roads, site preparation and clearing, and then active mining.

(c) Minerals are usually found under a layer of rock, known as waste rock or overburden; in a mining project, the ratio of extraction of ore to the generation of overburden and waste rock can be sometimes around 1:10.

(d) After the extraction of overburden, the next step is the extraction of the ore. Extraction of ore requires the use of huge equipment and machinery, specifically used for mining such as loaders, haulers, and dump trucks, which transport the ore to the mill through haul roads. The ore that contains high levels of metals also generates high levels of waste contain high levels of metals, for example, the ratio of copper content to waste in a good grade copper ore sometimes is only 1:4 and similarly, the ratio of gold content to waste of a good grade gold ore may sometimes be only 1:100 or more.

(e) After the ore reaches the process plant, the next step in mining is grinding (or milling) the ore. Here, the metal is separated from the nonmetal material of the ore and this process is done through a method called beneficiation and, in the end, the metallic parts are ground to fine particles for better extraction of metal.

(f) However, the milling process leads to the release of contaminants in the form of tailings which are the remains of the milling process of ore after the extraction of the metal which is of value.
(g) The final stage is the reclamation process when active mining stops, the site is reclaimed, and the mine is closed. Reclamation requires that the site should be brought back to the condition it was in before the mining started.

1.3.3 Feasibility of a Mining Property

The feasibility study for a mining property constitutes of many people who need to evaluate different commodities for different types of mines and plants in different environments; however, the outcome needs to accomplish one factor the potential of the property to draw financial opportunities from the investments that are incurred to it (Bullock, n.d.). Some of the main outcomes must include the mineral resource determination, selection of the best mining method for the resources, conducting a market analysis, understanding the infrastructure requirements, putting numerical measures to environmental and socioeconomic impacts, estimating costs for all the factors above, and portraying the accurate economic analysis of expected revenues and costs to determine the profitability of the project (Bullock, n.d.). The objective of the feasibility study should be to maximize the value of the mining property which assists in the decision-making for the company to develop and run it or to sell it.

The most recommended approach to feasibility studies as embraced by many large mining organizations contains three phases: the preliminary feasibility phase, intermediate feasibility phase, and the final feasibility phase (Bullock, n.d.). The preliminary study must include information and description collected through exploration in terms of location and access, surface feature, mineral-deposit geology, exploration activities review, geologic resource material tabulation, resource calculation method, land and water position, ownership and royalty conditions, history of the property, rock quality designation values, results of special studies conducted exploration department, report on any problems with local communities, etc. (Bullock, n.d.). This study should also give alternatives for mining and processing, which helps in assigning costs to the alternatives when observed by experienced mining and mineral processing personnel (Miguel, 1996).

The intermediate feasibility study phase is the continuation of the preliminary feasibility phase if this phase shows that the project can achieve the goals and objectives set by the organization (Bullock, n.d.). The goal of the intermediate feasibility study is to optimize each component of the mine and process plant while thoroughly checking all the project parameters of the preliminary study. The topography maps of the specific geographic location should be available in this stage and further exploration if required must be conducted and the results must be added to the current reports. The mine planning can also be started in this phase and two or three mining method alternatives should be finalized for comparisons later to select the final method of mining after carefully considering factors like safety and cost-effectiveness. The outcome of the intermediate study will assist in realizing economic viability and defining further testing requirements (Bullock, n.d.).

The third and final phase is the final feasibility study, which is a continuation of the intermediate feasibility phase if the results of this phase show definitive potential in achieving the company goals (Bullock, n.d.). The objective of this phase is to optimize the return on future investment. This phase assesses the impact of the final project features on various stakeholders like investors, communities, nongovernmental organizations, etc. Apart from other factors, the best alternative for mines and plants' final environmental impacts and final capital costs will be refined in this phase. Final presentations to the management will be made after which approvals as per the outcomes to continue to the next phase of project development will be received (Bullock, n.d.).

It may take up to 10 years or more for the exploration and development phase from the initial feasibility study to the production stage (ICMM, n.d.). The initial exploration stage involves highly advanced methods that include geochemical analysis of soils, airborne surveys to measure magnetic, gravitational, and electromagnetic fields, etc. The outcome of these tasks determines if the geographical location has sufficient deposits for mining. This is continued with the drilling of rock samples that are sent for testing to laboratories, which help in measuring mineral resources and ore reserves (ICMM, n.d.).

1.3.4 Block Modeling in LSOPM

In open-pit mining, at the feasibility stage, a complete pit design needs to be produced, while in the operation stage, the development of the pit should be such that it can respond to changes in metal prices, costs, ore reserves, and wall slopes. In the end, when the life of mine is completed the final pit design should be able to commence the adequate termination of a project in an economic way (Caccetta & Hill, 2003). The volume of the reserves is divided into fixed mining units for which the tonnage and metal content of each mining unit is optimized, and a spatial model of the data is created. The mining units are divided into millions of blocks and with the help of the description of the value of block production, scheduling, and optimization are done (Asad & Dimitrakopoulos, 2012).

Before the beginning of mining operations, the volume of the ore deposit is divided into blocks. The value of the ore contained in each block is estimated by using geological information from drill holes (Amankwah et al., 2014). A block model divides the entire ore deposit along with waste into blocks adjacent to each other as shown in Fig. 1.5 (Downing & Mills, 2017). The model can contain several hundred thousand blocks depending on the size of the deposit and the size of the blocks (Sevim & Lei, 1998). The cut-off grade determines the classification of the blocks in the block model according to which the various decisions in the mining sequence are undertaken (Moosavi et al., 2014). In the block model, the ore is distinguished

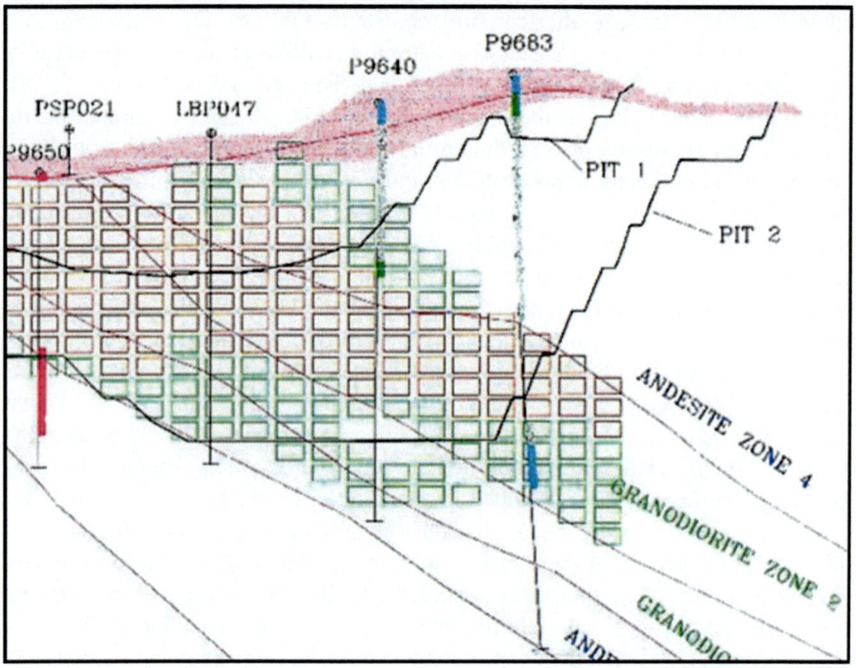

Fig. 1.5 A typical section for the proposed open-pit of a copper mine

from waste by establishing a cut-off grade, which is a function of commodity price, mining cost, and milling cost. As an example, a block with an average ore grade of 1% copper and 110 lb per cubic foot density will contain 11,000 tonnes of mineral, from which only 110 tonnes is expected to be copper (Amankwah et al., 2014). A profit is associated with each block of the mine after the estimation of the cost of mining and processing of each block (Amankwah et al., 2014).

The blocks with an average grade more than the cut-off grade are considered as ore blocks and the ones with an average grade less than the cut-off grade are considered as waste blocks (Sevim & Lei, 1998). Blocks are classified as ore blocks when the block processed has more revenue than its processing costs, and in case of waste blocks, the processing costs are more than the revenue (Lamghari et al., 2013). After the mining of the ore blocks, the material is sent to the mill for processing but to reach the ore block sometimes waste blocks are mined and the materials from the waste blocks are sent to the waste dump. However, sometimes, the ore blocks are not mined because many waste blocks need to be mined to reach the ore block (Sevim & Lei, 1998).

A basic problem in open-pit mine planning is deciding on which blocks to mine or not to mine, which is described as the problem of finding an open-pit mine design (Amankwah et al., 2014). Determining the optimum ultimate pit limit of a mine is the contour resulting from extracting the volume of material. The shape of the mine at the end of its life is given by the ultimate pit limit and optimum pit design is extremely important in all stages of an open-pit mine (Caccetta & Hill, 2003). In open-pit mining, when designing an optimal pit, the entire volume as described earlier is subdivided into blocks and the value of each block is estimated using geological information where for each block, the value of the ore less the excavating cost for the block is derived during the pit-design stage (Hochbaum & Chen, 2000). The optimum pit needs to be continuously monitored at all stages because the optimum pit and mine planning are dynamic concepts. The ore body is divided into fixed-size blocks and the block dimensions depend on the physical characteristics of the mine (Caccetta & Hill, 2003).

Three important factors during the determination of the geological block model are: whether a given block in the model should be mined or not; if so, when the mining should be done; and after mining the block, what processing technique should be used (Dagdelen, 2001). The answer to the questions above, when combined with the whole block model, helps in understanding the annual progression of the pit surface and the yearly cash flows that will take place during the mining operations. The decision taken during the planning stages of the block models has a huge long-term effect and is linked to the overall economics of the project (Dagdelen, 2001).

An important factor in determining returns on investments of millions of dollars is the production scheduling of open-pit mining operations, which poses a huge challenge for mining companies (Lamghari et al., 2013). The production scheduling method starts with the representation of the mineral deposit in the form of a three-dimensional array of blocks as explained earlier. The scheduling of production should take into consideration the proper order of which blocks to be removed at what stage from the ore body. The mining sequence, which is determining which blocks to extract and when to extract, is also part of the production schedule (Lamghari et al., 2013).

The production target set in the annual mine plan is achieved by the mine operations plan, which normally would include a daily, weekly, and monthly production schedule (Roman, 1999). Mine planning involves long-term planning and has processes that identify the ultimate pit-limit or overall extent of extraction and optimal production scheduling of the blocks (Asad & Dimitrakopoulos, 2012). In the production phase and to get the optimal ultimate pit limit, dedicated human and physical resources are required over several periods. To determine how much of the ore and waste must be moved to meet the production targets, the grade and volume requirements must be established (Roman, 1999).

1.3.5 LSOPM Processes

Open-pit mining is the removal of the soil from the ground, layer by layer, and this process is also known as the extraction process or mining of the ore in which the minerals need to be separated for further processing (CATAPA, 2016).

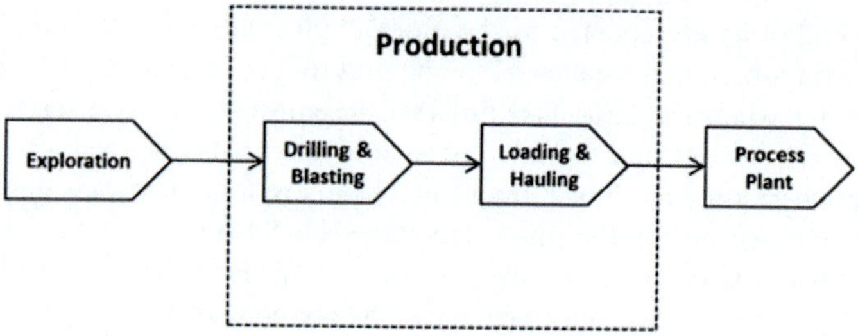

Fig. 1.6 Main operational areas of mining

The main operational areas of mining are given in Fig. 1.6. It starts with the exploration stage already described in Sect. 1.3.2. During production, there are two important stages, the drilling and blasting phase—concerned with rock breakage—when the ore is removed from the surface, and the loading and hauling phase—concerned with material handling—when the ore is transported to another place in large equipment. Finally, the processing stage when the ore is processed (Hartman & Mutmansky, 2002).

To commence a mining operation, it is a prerequisite that the existing reserves are economic, and this can be known from the results of the drilling during exploration and the interpretation of the geological landscape of the ore reserves (Asad & Dimitrakopoulos, 2012). The tonnes and grade of ore mined and the recovery of the contained metal in the milling and refining process are the factors that constitute the product volume. When the ore is too deep, too far away, or too small for development, it becomes uneconomical to develop and mine (McIlroy, 1999).

The grades of mining can be categorized into four distinct types: cut-off grade, geological grade, recoverable grade, and concentrate grade. The four types of grades (McIlroy, 1999) are:

1. **Cut-off grade** is related to the actual size of a deposit in terms of tonnes of reserves and the amount of contained metal. The cut-off grade is the margin beyond which it is not economic to consider production.
2. After the application of a cut-off grade but before the application of any mining plan, the amount of metal present per tonne of reserves is called the **geological grade**.
3. The amount of metal recovered per tonne of reserves that is mined and processed is the **recoverable grade.**
4. The percentage of metal contained in the ore produced from the mine and processed in the process plant is the **concentrate grade.**

In open-pit mining, the cut-off grade is an especially important factor that affects the life of the mining project (Moosavi et al., 2014). Ore and waste can be separated by using the cut-off grade factor, where if the grade of the material in the deposit is above or below cut-off grade, it is categorized as ore or waste, respectively (Asad & Topal, 2011). The cut-off grade is used to differentiate between ore and waste at the time of scheduling and most open-pit mines are designed and scheduled according to the cut-off grades calculated by using the method of breakeven economic analysis (Dagdelen, 2001).

The quantity of material mined and processed, and the product produced in the refinery can be indicated with the information of the cut-off grade. Thus, the optimization of the cut-off grade is vital for any open-pit mining project (Moosavi et al., 2014). The cut-off grade optimization can further be described as "an interactive process, including considerations such as metal price, mining, milling cost, the capacity of the processing plant, mining capacity in the mining operation, mining sequence, grade distribution of the deposit, and resulting cash flow" (Moosavi et al., 2014).

Large excavators and haul trucks are used to extract and transport the material to a processing plant or to an intermediate site which can be a mill, a leach pad, a stockpile. The material may be transported to a waste dump, depending on the expected profitability and processing-plant capacity (Espinoza et al., 2012). The initial capital expenditure decisions on the purchase of equipment like haul trucks, loaders, and processing plants and the development of infrastructure in the form of roads allow in the estimation of the rate at which the material is excavated and processed. Production constraints govern the factors for the rate at which material can be extracted from the deposit while processing constraints govern the factors for the rate at which it can be sent through a processing plant (Espinoza et al., 2012).

Chemicals like cyanide in the case of gold, or sulfuric acid for copper or nickel are reagents used for processing and when the mineral atoms bind with the chemicals, it forms complexes and the mineral is extracted from these complexes by using huge amounts of water (CATAPA, 2016). A recovery percentage is a percentage of total metal in the ore recovered, for example, a recovery of 92% means that 92% of the metal is recovered while the remaining 8% is lost in tailings (McIlroy, 1999). Mining and processing are two standard processes where the blocks are mined from the ground and then processed in a mill (Lamghari et al., 2013).

The excavated ore is sent to the crushers where it is crushed, split, and scraped into smaller particles which are kept in large heaps and are sprayed with a chemical solution and this process is known as heap leaching (CATAPA, 2016). The ore is sent to the processing plant for the metal content's crushing, grinding, and concentrating, and the product of the processing plant is called concentrate, which is fed to the refinery to produce refined metal (Asad & Topal, 2011). The processed ore is sold on the spot market or more often to companies having long-term contracts with the mining companies, and waste is left in piles, which must ultimately be reclaimed when the deposit is closed (Espinoza et al., 2012).

1.3.6 Mineral Processing

As per many classifications, there are over 4,000 different types of minerals, and minerals and metals are used together in the context of mining (ICMM, n.d.). The common group of metals as categorized in Table 1.10 in the appendix section at the end of the chapter. The ore extracted from mining operations along with the mineral mined contains other waste material rock, gangue, and uneconomic minerals (ICMM, n.d.). This ore must be treated and processed to produce an economically viable mineral product and the process of extraction, separation, and concentrating of ores to form an economically viable product is called mineral processing (ICMM, n.d.). The concentrated ore is then subjected to refining and smelting to produce metals of various grades of purity. The entire process of conversion of ore mined from a mining operation to a concentrated ore concentrate or commercially valuable minerals or separating the metal from the ore is done in a processing plant (ICMM, n.d.).

The processing/milling activities include adding chemicals to separate ore from waste, the ore goes through various phases of processing while the waste is pumped to the tailing dams. An alternative process in milling is the gravity separation technique where two or more minerals are separated through centrifugal forces, magnetic forces, or buoyant forces (ICMM, n.d.). The metal content after the separation process must be removed and refined and one common method used for it is smelting involving chemical breakdown of the minerals through heating and melting which are done in the smelter, sometimes specifically designed for minerals. The resultant concentrate is then sent to refineries for the purification process to purify it as per world standards which sometimes entails 99.9% purity. The process can also be done through hydrometallurgy in which the minerals are separated by dissolving them in certain acids or cyanide and then removed from it using processes like solvent extraction and electrowinning (ICMM, n.d.). The solids, liquids, and gaseous wastes mainly the airborne gases and poisonous minerals like mercury emitted to the environment that is the outcome of the refining process are major environmental challenges that if not handled with care can result in a major environmental disaster (ICMM, n.d.).

Case Study Outcomes

Mineral processing can be broadly classified into four stages or unit operations:

Stage 1: Comminution—This is the process of particle size reduction of the ore by crushing and grinding.

Stage 2: Sizing—The process of separation of various sizes of the crushed or ground ore by screening or classification.

Stage 3: Concentration—In this stage, the screened and classified ore is concentrated by making use of the physio–mechanical–chemical properties of the crushed, ground, and sized ore. The product of concentration is ore concentrate and tailings.

Stage 4: Dewatering—In this stage, the concentrated ore is separated into solids, liquids, and pulp by filtration, thickening, and drying.

The process plant operates various processes customized and designed specifically to separate and concentrate various ore types to obtain the planned concentrated grade and recovery economically. Thoroughly engineered and designed process plant equipment and instrumentation are principal drivers to achieve the designed concentrate grade and recovery for a viable and sustainable processing operation. Safety, productivity, and costs can be further optimized to enhance plant performance by various configurations of equipment, their designs, metallurgical and ore characterization studies, simulations, and by trained and skilled manpower to operate and maintain plant equipment and process parameters.

1.3.7 Understanding Ore Deposits and Optimizing Mine Design for Production

Ore bodies are not homogeneous (ICMM, n.d.). Plans to exploit the ore body to its maximum potential are possible when the ore body is defined well. While designing the mine, it is important to understand how big, what shape, and where the hole in the ground should be made considering the strength of the ground. The mining should be done in a way that the minimum amount of waste is extracted. The steeper the walls of an open-pit mine the less waste are excavated to expose the ore rock (ICMM, n.d.).

Case Study Outcomes

There is a requirement to understand the effect of a decision in one area over others which is why it is important to go over the details like the quality of the ore, the by-products, the best mining sequence that could be followed, and the cost that follows due to that decision. For example, in terms of gold mining, if there are two types of ore bodies, one is having higher manganese content in the ore and the second one is having lower manganese content in the ore and as per the contract with the customer, they accept only a certain percentage of manganese which is known as the injection limit. If the injection limit, in this case, is 1.5% or less, then for any additional quantity of manganese, the company will be penalized by the customer. This means that it is not just about making production targets but more about every other factor that can affect other departments in terms of cost or even the product.

During the sampling process, the samples taken from the drill holes let the mine planners know the ratio of ore in the samples and they differ from place to place. The sample data are entered into a database and the geotechnical engineer then models the geometry of the ore body, variability of the rocks, grade of ore in the rock how that varies and the strength of the rock, how that varies, the hardness of the rock, how that varies. Geostatistics is the branch of mathematics and is also used extensively in other disciplines where not only just the variable is important but also the location of the variable. This area of discipline is used in environmental work say contamination area and also used in meteorology, air pressure and temperature, and wind currents.

New techniques are required in the early stages of exploration to understand the ore body. One such example is X-ray diffraction, which is a technique that can be used to determine some of the physical characteristics where the material is bombarded with intense X-rays and the scan of the X-rays is used to understand the proportion of ore in the sample. This data will immediately be transferred to the geological modeling software to determine based on the shape and the size, the richness, and the proportion of gold in the ore body. Specialized software is used by the geologists to build a model of the ore body, which is then used by the mining engineer to design the pit to get that ore body out of the ground.

The initial planning stages are extremely important, for example, for a simple operation in a mine which is waste dumping, while mining from a certain area, a waste block is taken and dumped on the waste dump location. The sequential dumping will itself allow assessing all the opportunities in terms of dumping from east side to west side or from north to south, etc., and each option will give a certain cost and that will directly be linked with the sequence of mining relating to optimizing haul roads or optimizing waste dumping sequence.

Geotechnical engineering helps in defining the steepest wall angle that can be used to safely mine. Drill holes are used to take samples of the ground, the pit will then be excavated in, and various analytical techniques are used to measure the compressive and the tensile strength of the rock. The rocks have different typologies, there could be fracturing and faulting within the rocks, zones, or planes of weakness in the rock. Water content is also looked into because the water will tend to exert an internal pressure in the rock and force it to fall apart, which makes it weaker and also water will tend to lubricate the joints and the fractures in the rock that makes it easier to fall apart and reduces the friction between the various places.

Different modeling techniques are then used to model the strength of the rock mass to determine the steepest slope that can be mined in and it is not necessarily a constant value. In some areas, mining can be done steeper than in some areas where mining must be flatter and so while designing a pit, this constraint should also be considered. Long-range mine planners design the hole in the ground after which they work out the rate of mining in terms of how many tonnes in a year should they mine to maximize the profit. For example, an open-pit that has 500 million tonnes of material in it or waste, the question that needs to be answered is should this be mined as 500 million tonnes in 1 year or should this be mined in 10 years or rather 20 years? The slower it is mined the lesser equipment is required making the capital costs lesser, which will also slower the extraction of the valuable component, the ore. So, this might lower costs, but it also has lower revenues and hence lower profits.

The geography of a mine deposit is a crucial factor in mining. For example, the type of ore body in terms of hard rock and soft rock generally can cause changes in time and difficulty in mining because the soft rock is easier to manage compared with hard rock. Hard rock requires dealing with issues in drilling and blasting, hard digging, and milling or processing. Another example is having a single pit or multiple pits to operate in the same area of operation and its distance from the process plant because of its effects on transportation time. The advantage of having multiple pits is that personnel and equipment on the site can be moved to another operating pit when drilling and blasting are being conducted in one pit increasing the manpower and equipment efficiency.

1.4 Conclusion

Due to its vast requirements as a very core raw material provider for most of the other industrial sectors, the mining sector is indispensable. Mining was considered as an activity with unsophisticated methods in the past, but today with high-end technologies in the form of computerized remote-control equipment and complex machinery it has evolved into a very sophisticated industry with highly trained personnel maintaining the highest standards of safety and efficiency.

Mining is considered as a field that includes numerous challenges. The advancement of technology, mechanization, and the rise in the customization of information systems is expected to produce the required changes (CRIRSCO, 2018). As an industry, mining depends on the confidence of investors and other stakeholders. Depleting mineral assets necessitates measures to be taken during the extraction process of the minerals. In a competitive environment with a global downturn, many businesses boost efficiencies to survive using cost-optimization and efficiency strategies resulting in new growth.

1.5 Summary

The mining industry is vital and plays the role of a game changer and a contributing factor in the growth process of many national economies (CRIRSCO, 2018). One such example is the mining technology service sector that specifically provides technology-based products in mineral exploration, production, and processing proving a contributing factor for the sustainable development of Australia's mineral resources. The mining industry experienced major changes during the twentieth century, especially moving to open-pit mining techniques from underground mining. By 2010, most of the mine operations in the world were open-pit, followed by underground mining, tailings, and other methods (ICMM, 2012a).

The rising demand for minerals (especially metals), a highly competitive market, and many environmental and social issues being recognized on a long-term basis require leading companies to upgrade to high-end technologies. Information availability in the mining scenario and the outcome of using the same to achieve operational excellence and swift decision-making processes are especially important factors. It is important to bridge the gap between the perceived notion and the reality of information technologies being used to enhance information availability. Chapter 2 deals with the detailed understanding of information technologies and its prominent contribution to information availability for organizations and more specifically to LSOPM operations.

Appendix: Additional Information About Metals and Minerals

See Tables 1.6, 1.7, 1.8, 1.9 and 1.10

Table 1.6. Symbols of prominent metals and minerals

Metal/Mineral	Symbol	Metal/Mineral	Symbol	Metal/Mineral	Symbol
Aluminum	Al	Lithium	Li	Gold	Au
Boron	B	Magnesium	Mg	Silver	Ag
Beryllium	Be	Manganese	Mn	Platinum	Pt
Bismuth	Bi	Molybdenum	Mo	Radium	Ra
Carbon	C	Neodymium	Nd	Antimony	Sb
Cobalt	Co	Neon	Ne	Silicon	Si
Chromium	Cr	Nickel	Ni	Tin	Sn
Copper	Cu	Nobelium	No	Titanium	Ti
Dysprosium	Dy	Lead	Pb	Uranium	U
Iron	Fe	Palladium	Pd	Vanadium	V
Germanium	Ge	Praseodymium	Pr	Zinc	Zn
Mercury	Hg	Polonium	Po	Zirconium	Zr

Table 1.7 Relevance of various minerals/metals for day-to-day life

Minerals/Metals	Description of use
Minerals used for health and hygiene	
Iron, Zinc, Copper, Chromium, Molybdenum, Cobalt, and Manganese	Essential for proper health maintenance of humans, plants, animals, and microorganisms in required dosages
Titanium and Aluminum	Used for lightweight prosthetics
Lead	Acts as a radiation shield used for protection from X-rays
Copper	Used as natural antimicrobial touch surfaces in hospitals to prevent the spread of infections
Zinc oxides and Titanium dioxide	Used as ingredients for sunscreens

(continued)

Table 1.7 (continued)

Minerals/Metals	Description of use
Neodymium and Dysprosium	Simple daily use of electronic toothbrushes contain more than 30 types of rare earth
Minerals used for security and access to water	
Zinc	Zinc fertilizers make crops healthier, resulting in higher crop yields, better income, and longer sustainability of the land, water, and air management
Copper and Nickel	Desalination plants use copper–nickel alloys for high-corrosion resistance
Aluminum	Used for packaging in many industries for drink cans, foil trays, and food containers
Iron, Chromium, Molybdenum, Manganese, and Nickel	A range of stainless steel products are used as machinery and equipment for various steps of food growing, processing, manufacturing, storage, and distribution. These types of machinery are robust, can be easily disinfected, and easily cleanable.
Minerals used for urban living and infrastructure	
Copper and Iron	Used in pipes and drainage systems that last for centuries
Iron and Nickel	60% of all nickel is used to make stainless steel, which is used for cost-effective architectural solutions for multifunctional buildings and infrastructures
Aluminum	Provides a lightweight and durable of cladding buildings that enable no maintenance for years
Copper	Reduces the amount of energy needed to power electronic appliances due to its characteristics of being an efficient conductor of heat and electricity
Copper, Silicon, Molybdenum, Beryllium, Germanium, Gallium, and Indium	All needed in some amounts for solar panels
Minerals used for information and communication technologies	
Aluminum	The anodized outside layer of aluminum can be thinner than paint provides the modern smartphone with a light but the resilient casing
Aluminum, Cobalt, and Lithium	Used for batteries in smartphones
Gold, Silver, Copper, Tungsten	Used for electrical connections on the phone and other electronic instruments. The type of metal depends on the need of the equipment.
Neodymium, Praseodymium, Dysprosium, & Iron	Used for higher sound from a small place, also used for the vibration feature of the phones

(continued)

Table 1.7 (continued)

Minerals/Metals	Description of use
Aluminum, Potassium, Indium, and Tin	Used for making the screen of a mobile phone
Minerals used for efficient and low carbon energy	
Neodymium, Iron, Boron, and Dysprosium	Used in the permanent magnet generators that many wind turbines contain now
Copper	Used from 400 kg to 4 tonnes per turbine depending on the technology
Lead, Lithium, Nickel, and Sodium	Used for battery energy storage that is used for integration of renewables and stabilization of the electricity grid
Molybdenum and Zinc	Increases the longevity and performance of the high-end steel, used to build the tower. If the zinc is thermally sprayed, it gives a corrosion protection of over 20 years
Minerals used for safe and efficient transportation-electric vehicle	
Lithium, Cobalt, and Nickel	Used to eliminate the tailpipe pollutants release when it is running contributing significantly to local pollution reduction
Aluminum and Magnesium	Reduces the weight of the chassis and other components in a car by 50% in comparison to more traditional metals like cast iron and steel
Iron, Manganese, Molybdenum, and Vanadium	When added to steel provides for a strong but lightweight frame for cars
Copper	About 15 kgs are used in cars for the functioning of essential components in a car like motor, wiring, radiator, connectors, brakes, and bearings
Dysprosium and Neodymium	Used in the motor generators for more efficient combustion engines

Table 1.8 MCI rank of top 25 countries in 2018 according to ICMM

Rank	Country	Rank	Country
1	Suriname	14	Armenia
2	Democratic Republic of Congo	15	Tajikistan
3	Guinea	16	Mongolia
4	Burkina Faso	17	Bolivia
5	Kyrgyz Republic	18	Senegal
6	Mali	19	Zimbabwe
7	Sierra Leone	20	Guyana
8	Liberia	21	Peru

(continued)

Table 1.8 (continued)

Rank	Country	Rank	Country
9	Ghana	22	Sudan
10	Uzbekistan	23	Mauritania
11	Namibia	24	Zambia
12	Madagascar	25	Dominican Republic
13	Botswana		

Table 1.9 Production value of metallic mineral (including coal) for the year 2016

Rank	Country	Production value (USD billion)	Rank	Country	Production value (USD billion)
1	China	626.3	11	Mexico	28.9
2	Australia	123.0	12	Peru	27.1
3	Russian Federation	91.5	13	Kazakhstan	18.6
4	USA	89.7	14	Turkey	17.2
5	India	77.0	15	Germany	15.8
6	South Africa	48.9	16	Poland	14.6
7	Indonesia	47.5	17	Colombia	10.1
8	Canada	39.4	18	Ukraine	9.9
9	Brazil	36.6	19	Finland	8.5
10	Chile	33.5	20	Democratic Republic of Congo	7.9

Table 1.10 Common group of metals

#	Group name	Description
1	Precious metals	Used in a range of applications like jewelry, electronics, catalytic converters in cars, etc. Example—gold, silver, platinum
2	Base metals	Used as basic building materials all around and are of low value. Example—copper, lead, zinc
3	Ferrous metals	Used to improve the properties of steel and are metals with high iron content. Example—chromium, cobalt, manganese, molybdenum
4	Nonferrous metals	Unrelated to steel making, these can have an overlap with base metals. The distinction is made in the context of its nonrelation to iron. Example—aluminum, copper, lead, magnesium, nickel, tin, zinc
5	Rare earth metals	Not rare in the earth but the extraction method is complex and difficult. Used in small volumes for manufacturing of glass, ceramics, glazes, magnets, lasers, television tubes, refining petroleum, etc. Example—uranium, scandium, yttrium, lanthanum, praseodymium, neodymium

(continued)

Table 1.10 (continued)

#	Group name	Description
6	Alloys	Made by mixing two or more metallic elements to form a new, unique s ubstance that has differing chemical and physical properties to its parts. Demanding industrial requirements, such as extreme temperature resistance, strength for high-pressure applications, fatigue resistance, weight reduction, or toughness, often in combination, have led to the development of a wide range of alloys. Example—all types of steels

References

Amankwah, H., Larsson, T., & Textorius, B. (2014, January 5). A maximum flow formulation of a multi-period open-pit mining problem. *Operations Research International Journal*, 1–10.

Asad, M. W., & Topal, E. (2011, November). Net present value maximization model for optimum cut-off grade policy of open pit mining operations. *The Journal of the Southern African Institute of Mining and Metallurgy, 111*, 741–750.

Asad, M., & Dimitrakopoulos, R. (2012, April 25). Implementing a parametric maximum flow algorithm for optimal open pit mine design under uncertain supply and demand. *Journal of the Operational Research Society*, 185–197.

Bullock, L. R. (n.d.). Mineral property feasibility studies. In L. R. Bullock, SME Mining Engineering Handbook (pp. 227–261).

Burt, R. (1991, May). The international diffusion of technology in the early modern period: the case of the British non-ferrous mining industry. *The Economic History Review, 44*(2), 249–271. doi:https://doi.org/10.2307/2598296.

Caccetta, L., & Hill, P. S. (2003). An application of branch and cut to open pit mine scheduling. *Journal of Global Optimization*, 349–365.

Calvo, G., Mudd, G., Valero, A., & Valero, A. (2016). Decreasing ore grades in global metallic mining: a theoretical issue or a global reality? *MDPI*, 1–14.

CATAPA. (2016). *Large-scale mining: CATAPA*. Retrieved from CATAPA Web site: https://catapa.be/en/miner%C3%ADa/t%C3%A9cnicas-de-miner%C3%ADa/large-scale-mining.

Coulson, M. (2012). The ancient world (from the beginning of 1066). In M. Coulson (Ed.), *The history of mining* (pp. 1–16). Hampshire: Harriman House Ltd.

CRIRSCO. (2018). *About us: CRIRSCO*. Retrieved from Committee for Mineral Reserves International Reporting Standards: www.crirsco.com.

Dagdelen, K. (2001). Open pit optimization–strategies for improving economics of mining projects through mine planning. *17th International Mining Congress and Exhibition of Turkey-IMCET2001*, (pp. 117–121).

Deneen, A. M., & Gross, A. C. (2009). World mining machinery. *Business Economics*, 169–176.

Downing, W. B., & Mills, C. (2017, January 1). *Quality Assurance/Quality Control For Acid Rock Drainage Studies*. Retrieved from Infomine Inc. Web site: https://technology.infomine.com/enviromine/ard/acid-base%20accounting/Quality.htm.

E&MJ. (2012, March). Chilean service companies become world class. *Engineering & Mining Journal*, 68–80.

E&MJ. (2016, September). Alrosa developing new open-pit diamond mine. *Engineering & Mining Journal*, 32.

ELAW. (2010). Overview of mining and its impacts. In ELAW, *Guidebook for Evaluating Mining Projects for EIAs* (1 ed., pp. 1–17). Eugene, Oregon: Environmental Law Alliance Worldwide.

Ericsson, M. (2012). *Mining technology—Trends and development*. Polinares. Retrieved from https://www.google.sr/url?sa=t&rct=j&q=total%20large%20open%20cast%20metal%20%26%20mineral%20mines%20in%20the%20world&source=web&cd=13&cad=rja&uact=8&ved=

0ahUKEwjjxNq7kZbLAhXH_R4KHeKlA6E4ChAWCCMwAg&url=http%3A%2F%2Fwww.polinares.eu%2Fdocs%2Fd2-1%2Fpolinare.

Ericsson, M., & Larsson, V. (2013, January 22). E & MJ's annual survey of global investment. *Engineering and Mining Journal*. Retrieved from https://www.e-mj.com/features/2541-e-mj-s-annual-survey-of-global-mining-investment.html.

Espinoza, D., Goycoolea, M., Moreno, E., & Newman, A. (2012, December 6). MineLib: A library of open pit mining problems. *Annals of Operations Research*, 93–114.

Fiscor, S. (2016a, July). Advances with large rotary Blasthole Drills. *Engineering & Mining Journal*, 26–29.

Graham, C., & Evans, V. (2007, August). *History of Mining*. Retrieved from Canadian Institute of Mining, Metallurgy, and Petroleum: https://www.cim.org/en/Publications-and-Technical-Resources/Publications/CIM-Magazine/2007/august/history/history-of-mining.aspx.

Hartman, L. H., & Mutmansky, M. J. (2002). Introduction to mining. In *Introductory mining engineering* (2 ed., pp. 1–21). New Jersey: John Wiley & Sons, Inc.

Hochbaum, S. D., & Chen, A. (2000). Performance analysis and best implementations of old and new algorithms for the open-pit mining problem. *Operations Research*, 894–914.

ICMM. (2012a). *The role of mining in national economies*. International Council on Mining & Metals. Retrieved from https://www.icmm.com/document/4440.

ICMM. (2012b). *Trends in the mining and metals industry.* International Council on Mining and Metals. Retrieved from https://www.icmm.com/document/4441.

ICMM. (2014, October). *The role of mining in national economies (2nd edition).* London: International Council on Mining & Metals. Retrieved from https://www.icmm.com/document/7950.

ICMM. (2015, September). *Minerals and metal management 2020: ICMM.* Retrieved from International Council on Mining & Metals: www.icmm.com.

ICMM. (2018). *Metals & minerals: ICMM*. Retrieved from International Council on Mining & Metals: www.icmm.com.

ICMM. (n.d.). *Mining contribution index*. Retrieved July 10, 2020, from International Council of Mining & Metals Web site: https://www.icmm.com/en-gb/society-and-the-economy/role-of-mining-in-national-economies/mining-contribution-index.

ICMM. (n.d.). *Social progress in mining-dependent countries.* Retrieved July 19, 2020, from International Council of Mining & Metals Web site: https://www.icmm.com/social-progress.

ICMM. (n.d.). *What are minerals & metals?* Retrieved July 21, 2020, from International Council of Mining & Metals Web site: https://www.icmm.com/en-gb/metals-and-minerals/producing-metals/what-are-minerals-metals.

ICMM. (n.d.). *Where and how does mining take place?* Retrieved July 21, 2020, from International Council of Mining & Metals Web site: https://www.icmm.com/en-gb/metals-and-minerals/producing-metals/where-and-how-does-mining-take-place.

InEight Inc. (2013, November 20). *NEWS*. Retrieved from HD Web site: https://harddollar.com/brief-history-mining/.

Kirsch, P., Viswanathan, D., LaBouchardiere, R., Shandro, J., & Jagals, P. (2012). Health Impacts extend from the life of a mine to the life of a community—Knowledge gaps. *Life-Of-Mine Conference*, (pp. 1–8). Brisbane.

Lamghari, A., Dimitrakopoulos, R., & Ferland, A. J. (2013, July 3). A variable neighborhood descent algorithm for the open-pit mine production scheduling problem with metal uncertainty. *Journal of the Operational Research Society*, 1305 –1314.

Macrotrends. (2020, July 10). *Gold prices—100 year historical chart.* Retrieved from MACROTRENDS Web site: https://www.macrotrends.net/1333/historical-gold-prices-100-year-chart.

McIlroy, A. R. (1999). *The return from exploration success: Relating economic quality to geological quality.* Ann Arbor: UMI Publishing.

Miguel, S. A. (1996). *Cost components and methods employed in a feasibility study for a typical open pit and underground mining operation.* Ann Arbor: UMI company.

Mikesell, R. F., & Whitney, J. W. (2016). An overview of the world mining industry. In R. F. Mikesell & J. W. Whitney (Eds.), *The world mining industry* (Vol. 11, pp. 1–21). New York, Routledge: Taylor & Francis Group.

Minas Gerais. (2012, August). *Engineering and Mining Journal, 213*(8), 94–99. Retrieved from https://search.proquest.com/business/docview/1069233313?accountid=150425.

Mining Journal. (2016, October 20). *Rio delivers mixed numbers: Mining Journal.* Retrieved from Mining Journal Web site: www.mining-journal.com.

Moosavi, E., Gholamnejad, J., Ataee-Pour, M., & Khorram, E. (2014). Optimal extraction sequence modeling for open pit mining operation considering the dynamic cutoff grade. *Mineral Resources Management, 30*(2), 173–186.

Net Industries. (2016). *Surface mining: Net industries.* Retrieved from Net Industries Web site: https://science.jrank.org/pages/4358/Mining-Surface-mining.html.

Nicholas, C. F. (1910, May). The wrongs and opportunities in mining investments. *Annals of the American Academy of Political and Social Science, 35*(3), 207–216. Retrieved from https://www.jstor.org/stable/1011188.

Roman, P. A. (1999). *Joint enhancement of mine operations and maintenance: A structured analysis approach.* Queen's University (Canada). Ann Arbor: ProQuest, UMI Dissertations Publishing. Retrieved from https://search.proquest.com/business/docview/304542219?accountid=150425.

Sevim, H., & Lei, D. D. (1998). The problem of production planning in open pit mines. *INFOR*, 1–12.

Snedeker, M. (1990). History of mining. In B. A. Kennedy (Ed.), *Surface Mining* (pp. 1–14). Society for Mining, Metallurgy, and Exploration, Inc.

Statista, Inc. (2016). *2015 Global list of top mining companies based on revenue: Statista Inc.* Retrieved from Statista, Inc. Web site: https://www.statista.com/statistics/272707/ranking-of-top-10-mining-companies-based-on-revenue/.

Sykes, J. P., Wright, J. P., & Trench, A. (2016). Discovery, supply, and demand: From Metals of Antiquity to critical metals. *Applied Earth Science, 125*(1), 3–20. https://doi.org/10.1080/03717453.2015.1122274

Tilton, J. E., & Guzman, J. I. (2016). *Mineral economics and policy.* New York: RFF Press.

Van Niekerk, G. (2013). From volume to value: Cost optimization in the mining sector. pp. 38–41. Retrieved from https://www.kpmg.com/Global/en/IssuesAndInsights/ArticlesPublications/mining-asset-lifecycle/Documents/from-volume-to-value-v2.pdf.

Wenco. (2015). *Fleet management: wenco international mining systems Ltd.* From Wenco International Mining Systems Ltd. Web site: www.wencomine.com.

Chapter 2
Integrated and Advanced Information Systems in LSOPM Operations

2.1 Introduction

Data produced by manufacturing companies throughout the product lifecycle are stored in information systems and most information is one way or the other related to the products and is created and stored in various systems for various functions (Giddaluru et al., 2013). Organizations can take advantage of their information resources and knowledge assets, and this can be done by remembering and applying their experiences (Dinh & Fillion, 2007). Information needs to be organized logically so that the product information that is already existing can be used in new product development and this, in turn, reduces the time and resources used, to find information, which in turn reduces the overall product cost (Giddaluru et al., 2013).

The processing power of information technologies is enormous, which helps in analyzing information faster and gives results in minimal (Ramrathan & Sibanda, 2017). Information systems have become very resourceful in providing swift information in abundance and its role is perceived to be a requirement for every organization (Donnelly et al., 1978). Information systems make the task of differentiating meaningless data and required data easier for the managers (Hicks & Gulliett, 1981). The information available to the management is used by them to reconfigure resources dynamically, both operationally and strategically, in the interests of sustaining corporate competitive advantage (Haslam, 2007).

Organizations sometimes face a lot of challenges in the implementation and functionality stages of information technologies (Dinh & Fillion, 2007). Information system development is related to modeling, specifying, and realizing the concepts from the business domain, which are called informational concepts. Sometimes, there is a lack of connection between the information systems/functions and there is a huge gap in finding information about an existing product (Giddaluru et al., 2013). Table 2.1 shows the spending of various sectors worldwide on information technologies in the years 2009 and 2010 (Chaffey & White, 2012).

H. Qudrat-Ullah et al., *Operational Sustainability in the Mining Industry*,
Asset Analytics, https://doi.org/10.1007/978-981-15-9027-6_2

Table 2.1 Information technology spending by industry vertical market, worldwide (in USD Millions)

Industry	Total IT spending 2010	Total IT spending 2009	2009–2010 growth (%)
Banking and securities	390,488	379,855	2.8
Communications media and services	392,506	378,750	3.6
Education	64,148	62,607	2.5
Healthcare	88,996	86,215	3.2
Insurance	159,926	156,573	2.1
Local and regional government	179,664	176,747	1.7
Manufacturing and natural resources	426,085	415,480	2.6
National and international government	244,410	235,086	4.0
Retail	146,239	142,420	2.7
Transportation	105,703	103,689	1.9
Utilities	125,583	119,927	4.7
Wholesale trade	83,315	81,196	2.6
Total	**2,407,063**	**2,338,544**	**2.9**

Due to the improper integration of their software applications which happens when organizations implement systems according to processes, sometimes, organizations are dissatisfied with their current systems (Baltzan & Welsh, 2015). Innovation in the information technology field is necessary for the easy-to-use IT platforms, which will enable customers and staff of an organization to easily adapt to changes. Organizations that do not keep up with changes in new technologies are considered old-fashioned and unresponsive to market changes. The employees who resist change in the form of new developments in information technologies need to be persuaded to accept the changes in an organization to diffuse the changes efficiently into an organization (McCrohan & Harvey, 2008). Organizations today spend millions of dollars on information systems and do not give the required importance of training and educating the personnel who need to input data and keep up with data integrity (Lee et al., 2008).

Data integrity is an important factor and some of the factors that affect the integrity of data in an organization are a requirement of manual input of data to information systems, defective hardware, and unstable network systems and programs (Lee et al., 2008). Human error while inputting data is a factor that organizations should take into consideration while implementing advanced information technologies. Information system used in an organization depends on many complicated issues like the use of the system is mandatory, the effect of the experience of the user has on usage, the influence of workable information alternatives, and the limit at which the retrieved information is used (Whyte & Bytheway, 1995). This chapter is further divided into four sections:

- IT for knowledge management and business intelligence
- Interrelationship of IT and productivity improvements in LSOPM operations
- Interrelationship of IT and decision-making in LSOPM operations
- Current and future application areas for IT in LSOPM operations

2.2 IT for Knowledge Management and Business Intelligence

There are three important drivers for managing business information in an organization—information, people, and technology resources as shown in Fig. 2.1 (Chaffey & White, 2012). Data can be defined as discrete, objective facts without context and interpretation; information consists of data with added value and having context and understanding of a subject, thus forming a basis for knowledge; and knowledge is the combination of data and information. In the same context, an immensely popular term often used is knowledge management, it can be defined in terms of the acquisition, analysis, preservation, and application of knowledge (Dinh & Fillion, 2007).

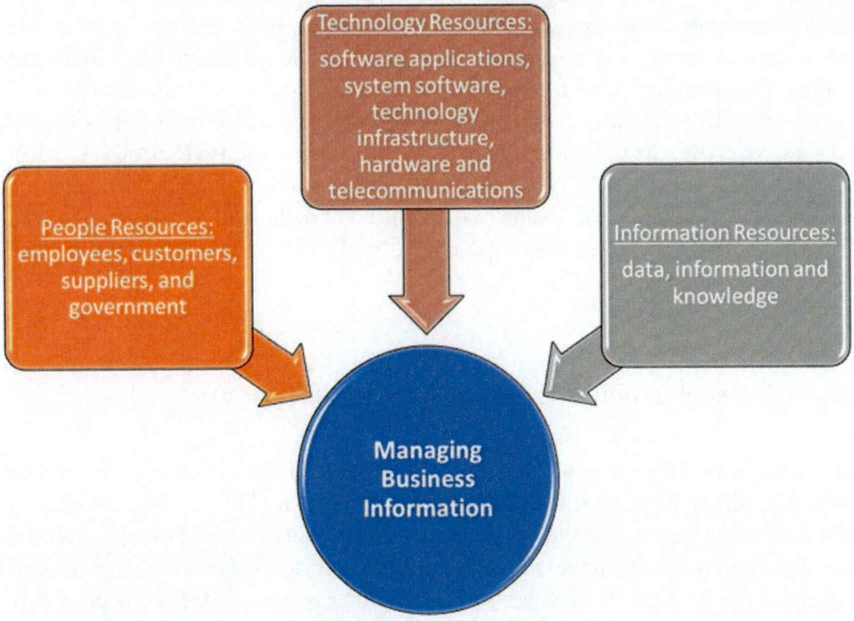

Fig. 2.1 Managing business information

Knowledge is built from data and initially processed into information (Uden & He, 2016).

Information can become knowledge when it is validated either individually or collectively as useful to be implemented in the system (Uden & He, 2016). In other words, knowledge is the process of obtaining data continuously and refining it. The definition of knowledge comes from the definition of data and information which comes from experience, values, and expert insights (Lovrencic et al., 2017).

The information technology field is an ever-changing field and organizations struggle to keep up with it and sometimes they end up over-depending on the technologies and if strategies on managing these changes do not work out well, it leads to information burden (McCrohan & Harvey, 2008). Adequate information is considered good, but many times information may be too much, and processing it to make use of it may sometimes prove to be a challenge for managers (Hicks & Gulliett, 1981). Information technologies have aided in providing the humongous amount of information and the challenge is to find information that is higher in quality. The relationship between data and information is that the poorer the data quality is the poorer the quality of information. Information quality is linked to the profitability and performance of an organization (Naicker & Jairam-Owthar, 2017). Companies dealing with knowledge management have the capability to change real-time data to real-time knowledge (Uden & He, 2016). However, the one issue pertaining to it is the problem of keeping up with the information security of individual customers (Adams, 2017). The evolution of the Internet of Things (IoT) has multiplied the problem of information security. The use of the Internet of Things has given the capability to knowledge management to handle real-time updates (Adams, 2017).

Real-time data from the wireless network can be availed from sensors and embedded technology for making new real-time knowledge among customers and vendors. Organizations can use IoT for gathering data from sensors and optimize performance (Uden & He, 2016). But in this case, there is a problem of dealing with enormous data because of which organizations have now started to look up at big data as a solution. Big data can be characterized in terms of volume, variety, and velocity, which reflects tremendous amounts of data, different natures of data generated, and the high speed of data generated, respectively. Another important aspect is the value that refers to the actual use of data collected (Adams, 2017).

2.2.1 Current Technologies Used in Organizations

Traditionally, organizations use three major classes of information systems, which are TPS, DSS, and EIS. A transaction processing system (TPS) is the fundamental business system that serves the first-line level of management (clerks and analysts) in an organization, an example of which is a payroll system or an order-entry system (Baltzan & Welsh, 2015). Management information system (MIS) has been used a lot as a tool in the making of decisions and decision support systems (DSS) are a subset of MIS for intelligent decision making (Nowduri, 2010).

A DSS is an information system that can analyze organizational data and then presenting it in a way that helps the user to make effective and efficient business decisions (Nowduri, 2010). Often DSS is used by all levels of people in an organization. DSS assists the top-level management in making strategic decisions, the middle-level management in making a tactical decision, and the first-line supervisors to make day-to-day operational decisions (Marcus & Dam, 2015). Decision-making, therefore, in any organization is very crucial not only just for organizations but also for individuals who greatly depend on these decisions for their survival (Nowduri, 2010).

DSS combines data and information to help everyone in an organization whether they are managers, analysts, or other business professionals during the decision-making process (Baltzan & Welsh, 2015). For example, researchers at IBM built BostonCoach (a limousine transportation company), a mathematical algorithm for a custom dispatch decision support system that combines information about traffic conditions, weather, driver locations, and request for customer pickup and determines which cars to assign to which customers. After launching BostonCoach, which had a revenue increase of 20% (Baltzan & Welsh, 2015).

Executive information systems (EIS) is a special type of DSS that supports top-level executives in an organization, usually has information from external and internal data sources, helps mainly executive end users, contains highly summarized information, and often used more for strategic decisions (Baltzan & Welsh, 2015). An IKBS is an information system that is used for problem-solving and requires the technical and professional knowledge of the field it is being used in, for example, medicine, engineering, chemistry, law, and architecture. IKBS development requires knowledge acquisition in the form of gathering and organizing of the knowledge about the domain of the application, and knowledge representation in terms of expressing or coding the knowledge (Yamakawa, 1997).

IKBS and a traditional program are different from each other, as the former has an inference engine that has knowledge about a domain in a knowledge base and is separated from the reasoning mechanism of the system while a conventional computer program is bound up in control statements such as conditional statements, procedural calls, and loops. One of the most prominent examples of IKBS is an expert system that is a computer program simulating human behavior in an area of expertise (Yamakawa, 1997). Simulation models are software models that simulate the economic situation of a sector or enterprise and can be used for decision-making and policy development (Marcus & Dam, 2015).

Expert systems are computer systems that contain knowledge in the form of generally accepted facts, published information, and experience based on personal experience of the expert (Marcus & Dam, 2015). Expert systems are used to make decisions and many times are considered to replace the decision-makers. Expert systems have made a tremendous impact on the field of business: in manufacturing, financial sectors, management, and office work (Jayashankar, 1989).

Unlike many softwares, expert systems relate to human natural language and have the ability of a human expert in a specific domain (Jayashankar, 1989). Other types of IKBS are PSS, DSS, consultation systems, intelligent computer-aided instruction systems, and intelligent front ends (Yamakawa, 1997). Another important concept for businesses is cognitive computing, which is considered as the future since it allows computers to interact with a human-like human (Ramrathan & Sibanda, 2017). The latest advanced information technologies comprise of modern systems for an organization need in various areas of the business as shown in Fig. 2.2 (Kadiyala & Kleiner, 2005). Information management systems have become a part of all processes including fleet management systems where equipment does not require human intervention and work without drivers making mechanization so advanced that minimal effort is required to operate the equipment.

Fig. 2.2 Advanced information technologies

2.3 Interrelationship of IT and Productivity Improvements in LSOPM Operations

Mining is a crucial industrial sector that requires ore deposits, which are a natural form of certain elements of economic value in the earth's surface and can be recovered by innovative technologies available (Dold, 2008). Computerization or automation is not new to the mining industry and for around 50 years has been used in the mining sector for increasing efficiency, safety by avoiding hazardous environments for mine operators and improving the accuracy and reliability of gathering data and processing it (McNab & Garcia-Vasquez, 2011). As with any other industry, the mining industry also began to modernize over time with the use of management philosophies and advanced information technologies.

Blasting in the open pit requires rock to be broken into small pieces through explosives which are put in drill holes 50-feet deep (Barrick Gold Corporation, 2018). The magnitudes of explosives are different for different places and are according to the rock formation of the area of operation. Blasting and its timing can also be controlled digitally. The blast execution can be done with a nanosecond precision that leads to greater control, safety, and efficiency (Barrick Gold Corporation, 2018).

Processing was done mostly manually for metals like lead, zinc, and copper; although simple crushing devices were in use from the sixteenth century, in the end, the ore and waste were separated manually (Burt, 1991). The ore that reaches the plant requires processing, which requires information regarding the amount to be processed at times. Recovery and grade control are two factors that are important for mines to be profit-making companies and, traditionally, are done manually by process plant managers (Hartman & Mutmansky, 2002). Conventionally, data collection, storage, and retrieval were done manually, in simple pen and paper style, and later with the use of spreadsheets (Fiscor, 2010).

The economy of the USA and many other developing countries use recent management philosophies like lean, six sigma, and TPS, but in combination with systems for information and knowledge, a practice that enhances business process management systems and business architecture (Furterer, 2009). Mining companies are currently looking for better grade ore deposits than trying to find better ways to mine them, which is why more ore deposits are discovered, but the speed of technology in mining does not increase accordingly (Yudelman, 2006a). Demand for minerals and metals is increasing and this makes the mining companies focus on increasing production target thereby not focusing on improving the technologies in mining (Yudelman, 2006b).

The exhaustion of mineral resources and mineral prices can be in balance only if the progress in technology is given its due importance. Mining operations pose a unique challenge in terms of setting and to conduct it with proper health, safety and in an environment-friendly manner, innovation in extraction processes regarding advanced technologies has been long advocated by many countries (Yudelman, 2006b). Some software vendors promote their products as applicable to the mining scenario with minor modifications, which is not a good practice, and mining organizations must emphasize on custom-made products or specifically make for mining requirements (Roman, 1999).

Another problem with mining organizations is that the decision to go in for new technologies is taken to modernize their techniques without knowing if or how the technology can aid in increasing productivity. The data in a mining company are enormous and hard to handle. Caterpillar came up with a new system in the early 1990s—the onboard monitoring systems that monitor production and equipment conditions (Roman, 1999). The challenge with this technology is that it produces a large amount of data, which mining organizations find difficult to analyze and thus end up only archiving it instead of using it. Mining companies should first use existing technologies optimally before planning to invest in new technologies (Jordaan & Hendricks, 2009).

Even with all the problems in organizations to realize the proper use of information technologies, metal mining companies tend to invest more in advanced technologies like huge and heavy size equipment, dispatch systems, and high-precision equipment positioning (Peterson et al., 2001a). Modern information technologies have become very sophisticated and extremely specific to different areas in mining and can be used in different phases of exploration, mining (which includes mine planning and drilling), and processing. In the mining process, automation has been done to both the machinery and equipment and is continued in use for monitoring, control, and communications systems and planning and design tools (McNab & Garcia-Vasquez, 2011).

Changes in technology are driven by many industrial objectives such as optimizing the cost of production, increasing the productivity of labor and equipment, discovery of new reserves, and extension of the life of mines, which are already existing (Peterson et al., 2001a). Haulage and loading operations are dynamic and complex, which requires several pieces of equipment interacting in a 3D environment (Fiscor, 2016a). The loading and hauling process in open-pit mining requires prediction of the load-haul cycle and this can be done by using several methods. The Monte-Carlo technique is the most common method that uses computer simulation, aiming to simulate the step-by-step operations of the entire truck and loader fleet, however, this technique is also critiqued due to its inaccuracy (Chanda & Steven, 2010).

Mining technologies include equipment, machinery, and technology used in the mining operations for monitoring, controlling, communication systems, planning, and designing tools and services (Peterson et al., 2001a). The investment in technology has introduced many changes to the large-scale open-pit scenario such as the following: (a) drill rigs are accurately positioned without the support of survey reports by drill operators to drill holes according to the designed depths; (b) hovels and trucks work 24 h per day regardless of the weather conditions, and (c) operator ability and job understanding have improved greatly resulting in significant benefits (Jordaan & Hendricks, 2009).

The objective of automation is to improve productivity and get accuracy by controlling the behavior of systems that are dynamic and extract maximum efficiency from the process. The advancement in automation depends on the advancement of information and communication technologies like wireless Ethernet, sensing technologies, navigation, and imaging technologies, etc. (McNab & Garcia-Vasquez, 2011). Businesses also use computer applications, which can be described in the business sense as a business activity or process to which an information system can be usefully applied (Warr, 1990).

Information technologies aid in mine development and investment opportunities through the extension of the life of mine when the operating margins are growing smaller and the equipment costs are getting higher. Monitoring productivity is important because of the changes in mine operations due to expansion, the addition of new technologies, and variations in demand (Peterson et al., 2001a). Mining companies have been seeking technological solutions to reduce production costs, and through this process, test and/or integrate various solutions to mining operations. An example of a mining company that aggressively seeks technological innovations is one of the largest technologically advanced copper mines in North America, the Phelps Dodge Morenci open-pit mine (Carter, 2006).

Many companies today provide advanced technologies for equipment and systems to save costs. Australia is a country where mining thrives as an industry and many specialized technology companies in Australia provide systems that are efficient and keep up with the required cost cut. The enterprise system created by the Australian companies shows mining proficiency in project management and an estimated 60% of the mining companies around the world using specialized systems (Fiscor & Casteel, 2008). Large mining companies are using the power of information, connected devices, and smart machines to understand the depth of the processes in the mining operations maximize production (McRoberts, 2016).

Mining companies are quite different from other manufacturing companies and require software companies that can provide technologically innovative systems specifically designed for the mining environment (Fiscor & Casteel, 2008). The mining industry in Australia, specifically, depends heavily on technological innovation to improve productivity (E&MJ, 2012a). The enormous amount of information produced by mining companies during various processes needs to be sifted through by mining personnel to make use of it in important decision-making. The availability of this information on any type of devices like mobiles and tablets makes it easier for swift decision-making (Fiscor, 2015).

A new concept in mining is that of a connected mine, where the information technology and the operations technology are all interconnected (McRoberts, 2016). Many mines face the problem of integration among various technologies available and face challenges in information sharing. A well-connected mine integrates all available systems so that information is shared and used when required by mining personnel. Equipment and software should be modernized and standardized across the mining value chain enabling future technologies to be easily integrated into the current systems (McRoberts, 2016).

Production intelligence software helps in optimizing production by understanding the relationship between mining equipment, raw materials, ore, and people

(McRoberts, 2016). Modern predictive software helps operators to optimize equipment use and increase throughput. Efficiency in a connected mine can be achieved in the areas of meeting daily targets; variability in production; predicting production outcomes, bottlenecks, and equipment failures; performance of assets; monitoring the condition of equipment; and improving throughput (McRoberts, 2016).

In mine planning, it is believed that what is shown in the paper cannot be executed exactly in the pit, and gradually, over a specific period, the planned targets could fall behind thus hindering the scheduling tasks and affecting productivity. The advanced information technologies available today can be used by mine planners in doing these tasks with minimum human intervention (Fiscor, 2010). Barrick Gold has invested hugely in innovative software solutions and has built a software operations center for the Nevada operations (Barrick Gold Corporation, 2018). The company also invests in skilled specialists who are technical to develop high-end tools with inputs from mining operators. The key driver of production goals is operational efficiency. Data analysis, autonomous controls, remote operations, coordinated blasting are some of the important measures the mining industry needs to keep an eye in the twenty-first century (Barrick Gold Corporation, 2018).

Technology is becoming increasingly important for mining companies to adapt and keep up with the competitive advantage factor (PwC, 2017). Productivity is the number one operational opportunity in mining and being digital has potential benefits for mining companies (EY, 2017). Ernst & Young, a professional consulting service provider, conducted a survey of global mining leaders. The survey concluded that new thinking along with a better focus on generating value will enhance performance and information technology is the key to harness productivity opportunities in the mining sector. Information technology should be used as a strategic tool for enabling better-operating models and practices and it requires mining organizations to make the right technology investments (Thompson, 2015).

Technology plays a positive role in enhancing productivity by helping in resolving issues that used to take hours to be done very swiftly (Mining Global, 2015). There are other productivity gains due to automation like a decrease in maintenance cost and the distance between the two machines can be reduced thus increasing operational efficiency (McNab & Garcia-Vasquez, 2011). Autonomous technologies like driverless vehicles, remove the errors which take place due to human performance and increases the efficiency of equipment utilization by reducing machine idle time and consistent speed in driving which in turn helps in improving efficiency in fuel consumption (McNab & Garcia-Vasquez, 2011).

Mining companies across the world are pioneering toward implementing best business practices by using the latest proven information technology systems. A revolutionary new technology known as the Blockchain technology, which was previously only used by other manufacturing companies, is now being used in mining companies (PwC, 2015). The Blockchain technology is already used by diamond mining companies as a quality assurance measure to verify quality, ethical extraction, and authenticity (PwC, 2017).

To achieve operational excellence, it is not enough just to cut costs and improve ore grades, mining companies need to harness the full potential of innovated technologies like a simulation of scenarios, driverless equipment, and big data analytics suites (Wenco, 2017). Operational efficiency can be achieved by automation and integration of systems through the supply chain. Mining deals with many remote locations and new technologies and analytic solutions help to increase operational performance by getting information at the right time, thus helping better transparency of information in the mining organizations in real-time (Business First Magazine, 2017). Real-time data and better data analytics help in processing decisions faster, which in turn helps in productivity factors like equipment usage, understanding machine movements to maximize efficiency (Whyte et al., 2015).

One important benefit of real-time data is knowing the location and state of each piece of equipment in mining operations and whether it is operated according to the plan (Whyte et al., 2015). New advanced technologies, like the operational effectiveness analytics solution developed by IBM's Haifa Research Lab, address operational excellence in the mining industry by identifying defects and minimizing process variabilities and can be customized for any type of open-pit mining operation (Hanson, 2016).

Mining operations that are remote operations, if well connected and integrated, can also achieve operational excellence and one such example is the IROC in Australia, of the company BHP Bilton (McRoberts, 2016). The remote-access technology of the company offers experts the opportunity to work from a single location and provide support to the companies' branches worldwide, including remote monitoring of equipment and alerting on-site personnel about the important issue, or sometimes virtually logging in to address the issue. In many mining companies, the data already exist, but it has not been changed to information, which can be useful, and data need to be collected, analyzed, and shared to change it into important information (McRoberts, 2016). Figure 2.3 identifies the technologies to reduce costs in Canadian mining companies, as given in a report from Deloitte (Procom, 2012).

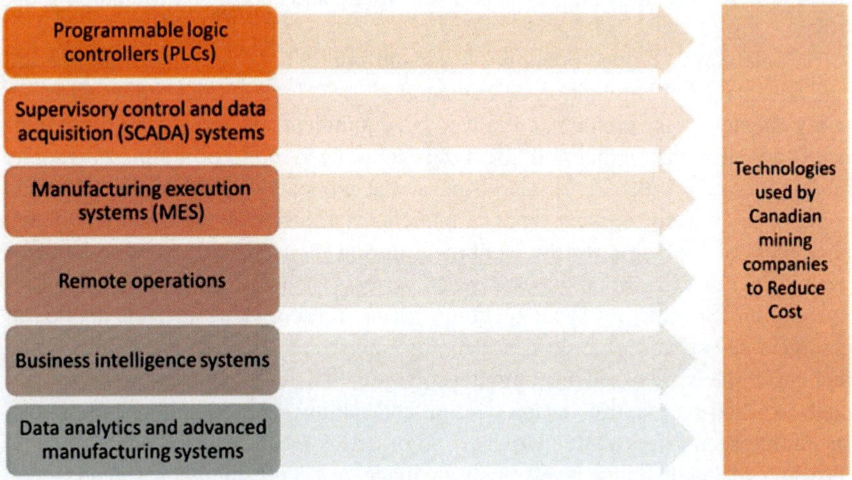

Fig. 2.3 Technologies for cost reduction in Canadian mining companies

Due to the growth and large investments in the mining sector, the dependency on mining equipment and machinery is huge. Equipment and machinery require maintenance and this, in turn, requires the manufacturers to come up with solutions that can integrate with the mining enterprise systems (E&MJ, 2012). Equipment utilization is maximized due to less downtime related to breaks and shift changes and the costs related to physical damage caused by human error, such as collisions and wear-and-tear of tires, are also removed (McNab & Garcia-Vasquez, 2011).

Technology can now help in analyzing the existing data and provide a set of data visualizing reports to show inefficiencies taking place in mining operations, which can be used by mining personnel to understand and rectify the inefficiencies. One of the large companies in information technologies, IBM, is using its full potential in analytical capabilities behind the mining companies to help increase production and make changes in the process to run it efficiently (Hanson, 2016). A breakthrough in productivity performance requires a change in the way of thinking about how mining works, which are where technological innovations play a pivotal role in changing the key aspects of mining (Whyte et al., 2015).

A report by Ernst & Young stated that the global mining sector faces many challenges or risks in the form of resource nationalism, worker shortages, and infrastructure access (Procom, 2012). Technology should be aligned to the business needs, which is not usually the case in most mining companies and thus the mining industry has a bad reputation for not delivering business value. Today, some companies run multiple instances of ERP systems across multiple departments; some types are single-user desktop software supporting technical tasks (PwC, 2015).

It is complex and expensive to acquire, implement, and maintain technology, and mining companies do realize this fact. Investments in technology add up to the operating cost, therefore, the mining companies must drive value from the investments (PwC, 2015). The companies can do cost-cutting by investing more in innovation in terms of automation in drilling systems, data analytics, and mobile technologies, and this, in turn, reduces the intensity in people, capital, and energy (Australian Mining, 2016).

The potential to reduce operating costs and improving operating discipline are some advantages of automation (Whyte et al., 2015). In open-pit mining, costs are mainly affected by the number and capacity of equipment, and the selection of equipment is a huge decision that will affect the economic viability of mining operations greatly (Bazzazi et al., 2009). Choosing a cost model for equipment selection is another challenge since many types of models exist, for example, traditionally, the major costs of looking after the equipment during its useful life were not considered and the modern cost model uses expert systems for decision-making in open-pit mine equipment selection or the technique of net present value analysis applied to equipment selection. Some cost models for equipment selection were done while other cost models are based on computer software, like the EQS that uses fuzzy logic which is based on maximizing production and minimizing the unit stripping cost. The mining sector uses a wide variety of information technologies to its advantage. It is also revealed that huge investments are made by mining companies on advanced information technologies (Bazzazi et al., 2009).

One of the critical drivers of economic growth and productivity is the communication flow, which is digital between countries, companies, and people. A survey of approximately 50 developed countries conducted by the World Bank showed results favoring the common belief that those firms that make use of modern information and communications technology have higher productivity rates and faster sales and employment growth (Kramer et al., 2007). Iteratively some studies also reflect that information technology in the form of internet technologies or specialized software technologies is responsible for higher productivity rates and better sales in many leading companies (Lewis & Steinberg, 2001; Rohleder et al., 2005; Tank, 2015).

Case-Study Outcomes

Productivity improvements have been related to factors such as employee efficiency, digitalization or use of information technologies, efficiency in time, and effective planning and designing. Information technologies are the second most important factor in improving productivity according to the study. The emphasis is largely on real-time information and the availability of information readily on the systems, which plays a particularly important role in improving productivity. Real-time information plays a vital role in enhancing productivity. People require to get online information and identify the mistakes or areas which can be improved and organize. Hence, online information from time-to-time from the operation affects and is also the key driver for cost optimization and productivity.

Mining companies need to invest in information technologies to achieve maximum efficiency of manpower and equipment. Mining requires the efficient use of resources for optimizing costs and increasing efficiencies. The change in productivity in one of the real-time examples included an increase of 10% in the first year and more than 14% in the second year in production with the same resources. These systems, in this case, helped in improving decision-making, cost optimization, and productivity. The most prominent effects of information technologies on productivity are shown in Fig. 2.4.

Optimization of time and cost are two other important effects that information technologies have on productivity. The real-time information availability has been linked to optimization of cost and time many-a-times, when the information is available it allows the managers to take a good decision sooner, reducing the time which

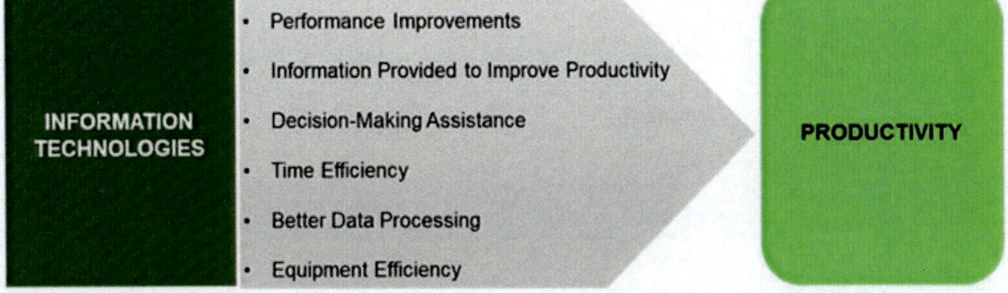

Fig. 2.4 Information technologies and productivity

in turn reduces cost. The influence of information technologies over productivity is shown in Fig. 2.5.

The inner circle shows the six most important effects of information technology over productivity and the outer circle shows the next nine effects of information technology over productivity. Equipment efficiency is a very important factor where information technologies are concerned with LSOPM operations. It has also been linked to real-time information availability on lead times, load-haul cycle times, utilization of equipment efficiently, and preventive maintenance along with the usual maintenance of the equipment. The performance improvement includes the improvement of the performance of the equipment along with manpower and the overall performance of the mining operations. A list of possible outcomes for the role of information technologies in improving productivity in LSOPM operations is given in Table 2.3 in the appendix section at the end of the chapter.

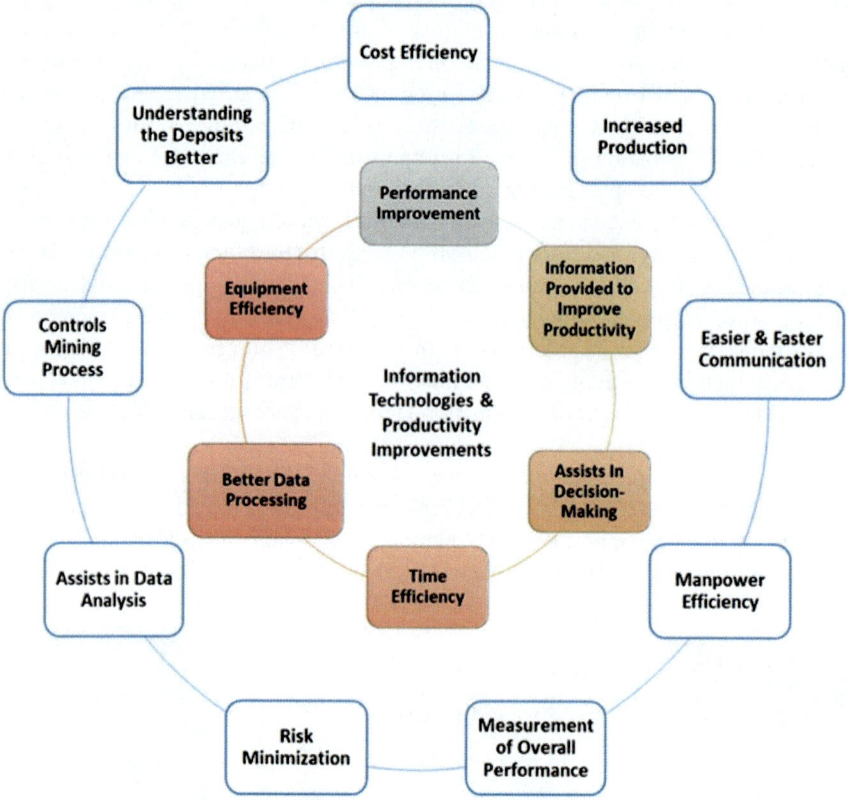

Fig. 2.5 Information technologies and productivity improvements

2.4 Interrelationship of IT and Decision-Making in LSOPM Operations

The success of an organization depends on many factors along with the way the decisions are handled, ensuring that the decisions are taken based on the knowledge possessed by the organization (Lovrencic et al., 2017). The key reason behind investments in information technologies is its ability to improving decision-making by providing access to more information. In the past, the decision-making style was more intuitive but now with the aid of information technologies, it has become more analytical (Ramrathan & Sibanda, 2017). Information technologies provide managers with the necessary data for making intelligent decisions (Donnelly et al., 1978). Information technologies play a vital role in logically making decisions and the swifter and more accurate the information the better results in terms of decision-making (Hicks & Gulliett, 1981). The information requirements depend on the level of the organization and the type of decisions being made (Donnelly et al., 1978).

The types of information can be classified according to the levels of the organization as the planning information required by the top-level management, the controlling information required by the middle-level managers, and the operational information required by the employees carrying out the day-to-day activities (Donnelly et al., 1978). Strategic planning is a continuous process and the main purpose of implementing strategic planning is controlling the organization. Information systems have many strategic impacts on an organization (Kadiyala & Kleiner, 2005):

(a) it helps to use resources optimally,
(b) it is known to reduce costs and identify potential market segments,
(c) it manages information effectively to simplify things, and
(d) it assists in satisfying the diversified needs of a business's customer base.

Strategic planning of the information technology requires that there is an alliance of the information architecture of the organization with the mission, vision, goals, and objectives of the organization. The information management applications can be categorized in terms of their support for planning and control at the strategic, tactical, or operational level and this is referred to as the Anthony model (Chaffey & White, 2012).

For planning and controlling to be effective, every organization requires relevant information, and the better the information the better is the result of the decision (Donnelly et al., 1978). Decision-making can be improved if the problem formulation and the solution phases are improved. One way of improving decision-making is through technology wherein computers and software have played an especially important role in recent years to make the decision-making process faster. Technology today helps managers to make group decisions without having face-to-face meetings and the real-time flow of information is helping hugely in reducing the time in decision-making (Hitt et al., 2014).

As an example, Hitt et al. (2014) discussed how the company "JetBlue uses technology to make many of its daily managerial decisions, which has helped JetBlue succeed while other airlines have failed" (p. 352). Airline companies lost $60 billion or more, cumulatively, between 2000 and 2009, and while information technology budgets were either cut or frozen by other airlines, in contrast, JetBlue nearly doubled it. JetBlue used workforce optimization technology called Blue Pumpkin and it is the first to create a paperless cockpit (Hitt et al., 2014). All pilots in JetBlue are equipped with laptops so that they can access flight manuals and make the pre-flight load and balance calculations, which saved JetBlue an estimated 4,800 man-hours a year. Each plane saved 15 min before take-off, which helped JetBlue add approximately 1,500 flights annually (Hitt et al., 2014).

Previously, due to the dearth of information people relied on manual processes for making decisions; but currently, the enormous quantity of information and understanding of the same information to make good decisions is growing exponentially (Baltzan & Welsh, 2015). It is inevitable to take the aid of information systems due to overly complex decisions that involve a huge volume of information that needs to be used to make fast decisions. However, dependence on information systems has its challenges.

Organizations invest extensively in information technologies but many a time, while the systems have helped to decrease transaction costs, they have been not been effective enough in improving decision-making (EY, 2013). Therefore, it is required to understand the systems and draw intelligence from the systems for effective decision-making.

Information systems have evolved from a simple database to classical business intelligence, and more currently, to the semantic business intelligence model. The semantic business intelligence model, unlike the classical business model, does not just help information from within an organization to be used, but also allows users in the organization to have information regarding the organization from the web (Airnei & Berta, 2012). New information technologies like Product-Service System (PSS) involve the consideration of factors that relate to product lifecycle, linking products and services closely, and making links with customers and other manufacturers to help in product/service delivery (Durugbo et al. 2011). Business process management systems are the collaboration of business and information systems and help in integrating all information technologies into one holistic system (Baltzan & Welsh, 2015).

The mining industry is quite different from any other industry and its challenges vary in terms of information technologies like for example the ERP systems in mining find it hard to keep up with many real-time aspects of mining. To make informed business decisions, it is important to have the right information, at the right time, to the right person. Right decisions can be made if the right information is gathered, analyzed, reported, and applied to the business (PwC, 2015). Mining companies usually have a large amount of data from drills, trucks, processing plants, etc., but in many cases, less than 1% of these data are used for any kind of decision-making. Investments in systems and tools help to build a foundation for better decision-making. The data provided by the systems are helpful in being used with decision-making algorithms to enhance better decision-making (Whyte et al., 2015).

In mining operations, the use of information technologies aids in decision-making across multiple job functions (McRoberts, 2016).

Some of the important decision-making factors carried out by mining personnel efficiently with the help of specialized technologies are (McRoberts, 2016):

(a) tracking equipment operators for overall equipment efficiency and mean time between failures,
(b) reduction in unexpected downtime occurrences done by maintenance technicians to efficiently monitor the asset health regularly and predictive maintenance,
(c) reviewing the ore grade and quality of the product by quality managers,
(d) viewing of crucial factors like real-time cost of production by site managers, and
(e) comparisons between real-time operations and commodity prices for adjustments to be done accordingly, done by top executives.

Recent technologies that are server-based solutions such as dispatch systems and production management systems can support decision-making in operations across mine sites and can increase productivity. Many mining companies do not integrate existing systems with modern systems and are not able to benefit from modern technologies (Thompson, 2015). To reduce the system integration gap that can improve decision-making, an end-to-end approach to be taken in mining organizations. Although the mining industry has embraced the new information technologies to bridge the digital disconnect between the systems, the rate of development is slow in comparison to the opportunities available (EY, 2017). It is, thus, evident that a detailed study on the contribution of information technologies in improving decision-making in large-scale open-pit mines is required.

Case Study Outcomes

Information technologies and decision-making are closely related to the amount of information available due to the information being gathered and analyzed by systems. The decisions require the availability of information faster for decisions to be taken in terms of equipment efficiency, manpower efficiency, and general productivity improvements. The study shows a strong relationship between efficiency in improving decisions due to the availability of information being faster with the aid of information technologies. Figure 2.6 shows the most prominent effects of information technologies on decision-making.

It is evident from the study that real-time monitoring of all movements in the information technologies helps the personnel to make decisions instantly. One such example is the Fleet Management System that requires intensive capital investment and is used to optimize the equipment fleet in a mining operation. If a piece of equipment is idle, or not working according to the benchmark or industry standards, or is not operated well by the operating personnel, or requires a preventive check-up in terms of maintenance, are scenarios during which a manager can decide various solutions for each scenario as guided by the systems in terms of options.

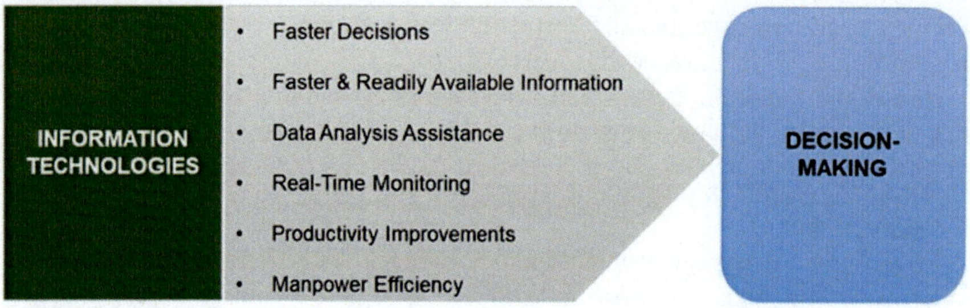

Fig. 2.6 Effect of information technologies on decision-making

In one of the real-world situations, when a manger is required to know which truck is running where at that moment and may be required to recall a truck. Previously, this could be done only after a shift change or shift completion but with the help of digitalization, this can be done any time within a few seconds, even if the manager is at home. Communication lines can be open to any truck drivers, any pump operator, any tailing dam operator from his/her home, and decisions can be made instantly that show the huge impact of IT over decision-making.

Information technologies also aid in setting up benchmark and goals, improving performance decisions, and maximizing production. A target achieved at a point of time can potentially be the benchmark for the next time, whether it is in terms of volumes of material extracted, the number of hours the equipment was used to extract that volume, or a number of hours manpower was used for the same. Figure 5.5b shows the inner circle in which six factors closely affecting decision-making when information technologies are used and the outer circle in which six other factors that affect decision-making in LSOPM operations when information technologies are used.

Information technologies should be used as an end-to-end approach by the mining organizations. Mining operations rely heavily on information technologies, one of the reasons being the large area of operations. Information technologies must be used efficiently, and many mining organizations are not using technologies very efficiently. The reason for the above conclusion is the readily available data in systems, the aid of information technologies in gathering and most cases analyzing data, and real-time monitoring of data which aids in improving productivity as shown in Fig. 2.7. The inner circle includes the factors most relevant effects of swift decision-making on productivity and the outer circle depicts the factors, which have a lesser effect on productivity in comparison to the inner circle factors. A list of possible outcomes for the role of information technologies in decision-making in LSOPM operations is given in Table 2.4 in the appendix section at the end of the chapter.

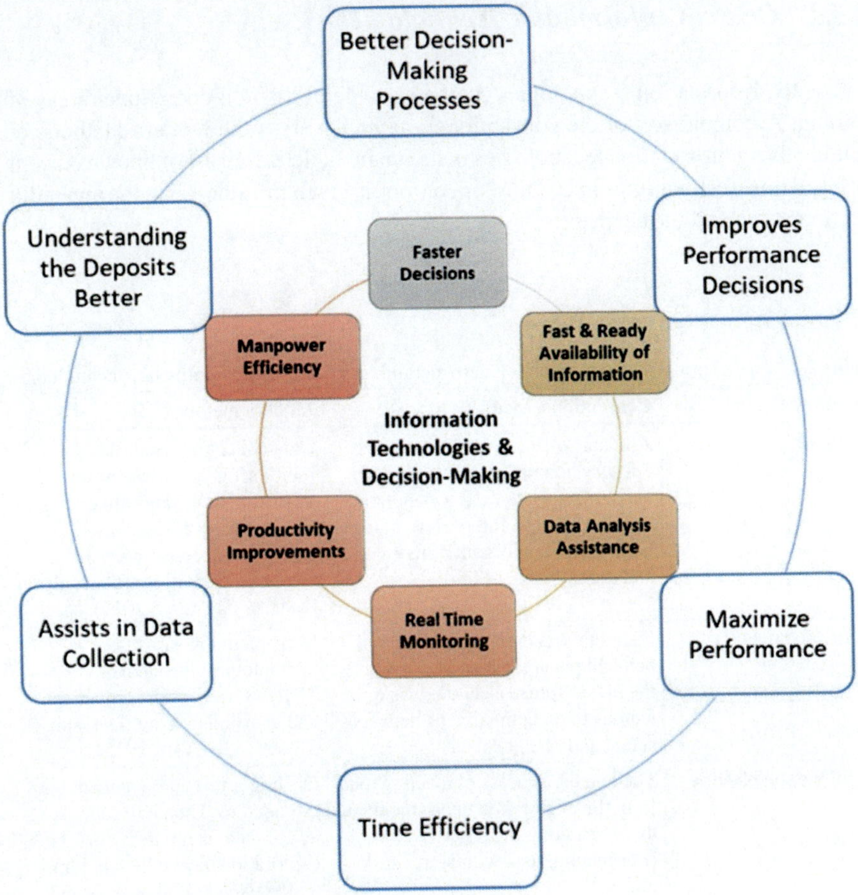

Fig. 2.7 Information technologies and decision-making

2.5 Current and Future Application Areas for IT in LSOPM Operations

Information technologies are becoming an overly critical element for mine operations to attain sustainability in mining. The mining industry became a part of the information technology revolution leading to advantages in mine design done using sophisticated computer systems, software for mine planning, and condition-based monitoring, along with rock analysis software, and geology mapping software (Roman, 1999;Yudelman, 2006b). Advanced maintenance technologies also include remote condition monitoring, which has the potential to identify failures prior to problems resulting in repairs to be scheduled prior to failures (Lewis & Steinberg, 2001).

2.5.1 Current Information Technologies

Currently, information technologies can be classified based on operational areas of mining. A comparison of the conventional methods versus the modern methods of mining using information technologies is shown in Table 2.2. A list of known current Information Technologies in LSOPM operations is given in Table 2.5 in the appendix section at the end of the chapter.

Table 2.2 Conventional methods versus modern methods of mining using information technologies

Areas of mining	Conventional method	Modern method
Exploration	All data collected, stored, and retrieved manually, in a traditional pen and paper style. It was not only difficult to store and retrieve data but it was also difficult to monitor and control it	Data collection, analysis, monitoring, and use of it for other functional areas like drilling using advanced information technologies like Geovia (GEMCOM) (Dassault Systems GEOVIA, 2015)
Mine planning	Planning was done in a traditional pen and paper style and later with the use of spreadsheets. It was quite tedious to do comparisons between actual and planned data	Mine planning software technologies like Geovia's SURPAC help with comparisons and decision making (Dassault Systems GEOVIA, 2015)
Drilling and blasting	The lengthy process of availing data from the exploration department and the time taking the process of coordinating data with actual drill points	Drilling information technologies provided by Thunderbird make use of data integrated with exploration data, which in turn, makes use of GPS systems and gets accurate results faster (Thunderbird Mining Systems, n.d.)
Loading and hauling	Even though manual loading and hauling were done hundreds of years ago, the equipment that replaced it still required manual reporting systems about availability, maintenance, and position information	Fleet management systems provided by WENCO, give real-time information about equipment status, position, and many extra details useful in fast decision making (Wenco, 2017)
Process plants	The ore, which reaches the plant, requires processing which requires information on how much to be processed and recovery and grade control were two factors that are important for mines to be profit-making companies. Manually, this was done by human experts and did take a long time	Expert Systems like e Scada provide with data that help in extremely fast decision-making (Fiscor, 2010)

- *IT in Mine Exploration*

GEOVIA (formerly known as GEMCOM), which uses high-end information technologies especially for exploration, comes in as a package having mining solutions like GEMS, Surpac, Minex, Whittle, and InSite (Dassault Systems GEOVIA, 2015). Datamine is the company providing mining software technology and services to mines requiring planning, managing, and optimizing mining operations and their systems help across the entire value chain from exploration, geo-statistics, resource modeling, mine planning to operations management (Datamine, 2015). Maptek is another leading company that provides innovative software, hardware, and services for the mining industry worldwide with a range of products, and its popular software, like Vulcan, is widely used in many large-scale mining companies (Maptek Pty Ltd., 2018).

- *IT in Mine Planning*

Mine planning software, MineSight from Mintec, USA, assists in every stage of mine planning in terms of the mineral deposit evaluation process, providing an economic planner with display tools for optimizing tasks, and detailing with the truck-haulage simulator and strategic planner (Fiscor, 2010). 3D block models of ore reserve calculations, blast planning, ore grade predictions, and mine design planning are created using software modeling packages like SURPAC, Vulcan, and Medsystem (Roman, 1999). Some companies develop advanced information technologies based on the mining requirements of a country.

Devex is Brazil's largest mining software house and its most popular software is SMARTMINE, which is used to control, operate, and optimize open-pit and underground mining operations. Canada's Toronto-based company, Quantec, though primarily not a software company, has developed new and innovative software technologies with 3D acquisition systems like Titan 24, which provides the mining companies with the ability to measure certain characteristics such as resistivity or conductivity of the earth (The service sector, 2012).

Pit optimizing is an important factor and specialized software is the key to achieving the optimum result. Australian mining IT company, Micromine, specializes in innovative software solutions that optimize the mining pit using input parameters like pit slope constraint, mining and processing costs, metallurgical recoveries, and selling price (Fiscor & Casteel, 2008).

- *IT in Drilling and Blasting*

The blast-hole drilling process is repetitive and is easily adaptable since the process is not overly complex and productivity improvements have been made with systems that have auto-drill functions (Fiscor, 2016a). An automated control system, which can be operated from a line of sight of about a mile away, offers high precision GPS and is in the form of a travel kit which can be placed in a pickup truck. BHP's mine automation readiness engineer elaborated that autonomous drilling led to increased machine utilization and consumable life (Fiscor & Casteel, 2008).

DEI™ from Thunderbird Mining Systems was introduced to the mining industry in 1982 and has been installed globally since then in many mines used mainly for the drilling process helps to increase the productivity, decrease overall drilling costs, and improve blasting efficiency in mines (Thunderbird Mining Systems, n.d.). The systems from DEI™ are implemented and used in mines globally, which include countries like Australia, Chile, Kazakhstan, Suriname, Ukraine, and the USA (Thunderbird Mining Systems, n.d.).

The materials-handling company, Transmin from Australia, developed a rock-breaker automated system, Rocklogic, which helps to reduce the size of material in crushing operations and provides rock-breakers with an integrated solution for remote operation, collision avoidance, and automated parking technology (E & MJ, 2012a).

The Canadian company Minalytix created software for the collection, management, and sharing of drill data in a simplified manner (Hiyate, 2017). The software from Minalytix requires only an internet connection to access the system and the system synchronizes any data entered on-field or office premises automatically (Hiyate, 2017). Leica Geosystems is an Australian-based software company that develops monitoring and navigating systems for drilling and blasting, and various equipment and machinery in surface mining. Leica Geosystems is a product suite of highly advanced technologies that can be used for increasing productivity and optimize operating costs (E & MJ, 2009).

- *IT in Loading and Hauling*

Two relevant Original Equipment Manufacturers (OEMs) in the mining industry providing mining technology systems are Caterpillar and Komatsu via Modular Mining Systems (Jordaan & Hendricks, 2009). Apart from that, there are non-OEM suppliers like Wenco and Jigsaw, which supply only technologies that integrate into the current information system of the mining companies. The decision to rely on OEM providing technologies or implementing technologies from non-OEM providers depends on the amount of risk involved in terms of capital, integration, etc. (Jordaan & Hendricks, 2009).

Fleet management systems provided by the Canadian company, Wenco International Mining Systems Ltd., for surface mining operations provide software and hardware for giving precise information on equipment activity, location, time, production, and maintenance (Wenco, 2017). Mining engineers use computer programs for the estimation of open-pit haulage truck requirements, like for example, Truck And Loader Productivity And Cost (TALPAC) and Caterpillar's Fleet Production and Cost (FPC) analysis (Chanda & Steven, 2010).

Automated drills and driverless technologies were developed in the 1960s and 1970s and the recent developments include operations that are autonomously requiring no human intervention, and for increasing efficiency and coordination automated components are being integrated (McNab & Garcia-Vasquez, 2011). The example of driverless trucks being run successfully in open-pit mines can be taken from the Rio Tinto mines at its Yandicoogina and Nammuldi mine sites (Diss, 2015). An operation that is at the center in Perth, around 1,200 km helps in controlling 22 driverless trucks remotely (Diss, 2015). Barrick Gold Corp. has begun its testing in surface autonomous haulage technologies in its mining operations in Nevada. The mine site is considered ideal for testing because currently it is in its mine development phase and core operations, therefore, will not get affected (Elko Daily, 2018).

- *IT in Process Plants*

Minerals can be tracked with the new technology of radiofrequency ore tracking system, which identifies ore or waste boosting efficiency while keeping costs down. Information about the material is sent to the database allowing mining personnel to handle the flow of ore and waste (Campbell, 2007). Multiple-criteria decision-making integrates common sense with analysis based on value-judgment, empirical data, normative data, quantitative data, and descriptive data (Pan et al., 2010). The MCDM approach supports advanced systems concepts, previously used for procedures in data management, modeling, etc., improvements in the decision-making process, and this method is used in the optimization of most of the technological systems today (Pan et al., 2010).

- *IT Real-Time Monitoring and Control*

While information gathering is important, it is also important to make effective use of the information, which is where real-time systems become a significant contributor to the assimilation of data (Peterson et al., 2001a). Recently developed GIS help mining organizations understand and manipulate spatial data and help analyze and develop scenarios based on initial mine design, operational plans, stockpile scheduling, equipment utilization, and expansion options. Dispatch system uses GPS to: (a) monitor mobile equipment positions; (b) directs trucks to the shovels, crushers, stockpiles, and dump-point; and (c) optimize equipment use and material flow in real-time (Peterson et al., 2001a).

Critical information technology requirements in mining are sensors used for real-time process monitoring and optimization, and integration, position monitoring, onboard computer hardware, advanced control algorithms, and wired and wireless communications (Peterson et al., 2001a). Mining operations need to be monitored in performance by the integration of data in terms of many factors like the scheduling of maintenance, several vehicles in the pit, haul distances, operating hours of the equipment, loads, size of the particle, and speed of processes (Peterson et al., 2001a). Another advanced technology used in the mining sector is the use of drones for aerial surveying and volume calculations (Hiyate, 2017).

Mining companies in Australia have mines in remote areas, globally, and when mining personnel travels, it is difficult to access data from the mine site (E & MJ, 2012a). MinePoint is an ERP software technology developed in the Microsoft Dynamic AX platform and offers mobile business software that enables mining personnel to access data from the mines anywhere at any time (E & MJ, 2012a). Mine-specific challenges are unique, and they can be overcome by using specialized technologies (Fiscor, 2015).

The Canadian company Naptec Technologies uses sensor technology and 3D real-time intelligence for securing all mine-related information (Fiscor, 2015). The Australian mining software company, Mincom, has developed IMS, which is a suite of technology products between real-time production control environment and the back-office transactional environment (Fiscor & Casteel, 2008).

The Swiss-based company, Endress + Hauser, combines state of the art technologies and specialized services to develop mining projects (E & MJ, 2012b). Pitram systems from Micromine record all data regarding mining activities in terms of equipment and processes and enable users to use the data for the long term. A new technological innovation in mining is virtualization, which is the virtual version of the mine's hardware. Administrative tasks and backup data are centralized in case of virtualization. Andes IT is a company in Chile that provides services in business processes through virtualization (E & MJ, 2012).

- *IT Real-Time Monitoring and Control*

The right analysis of data requires people who understand data analytic methods and interpret it correctly, which requires qualified data analysts to understand the gathered data and analyze it appropriately. In terms of safety, nowadays there are radar systems in mining (Baggaley, 2017). In a very big open pit, slopes are being monitored through data, so wherever there are any safety threats, continuous real-time monitoring helps in see how the wall of the pit behaves and in case of any movement which can be known by the threshold limit of the movement, the activities in the mining area can be withdrawn or stopped to ensure safety (Baggaley, 2017). The monitoring is still done manually in many mines even now where personnel must go to the pit, put the instrument, then take the measurement, and then decide whether to continue operations or not.

2.5.2 Types of Technologies Required in LSOPM Operations

Many times mining organizations do not assess and analyze well to implement the information systems as per the requirement of their businesses and end up with selecting technologies that may be modern and new but do not fit in completely with the business requirements. LSOPM operations are huge and require the use of general and specialized information technologies for day-to-day operations to sustain for the long-run and for achieving operational excellence.

Case Study Outcomes

Systems allow for better accountability as it is possible to keep tag cards on employees to know where they are at every point of time during their presence in the pit. This improves employee efficiency too as they are answerable to every minute that they are idle and not working in the field. It is required to first realize the full potential of the mine operations in every area and prepare a budgeted metric through key performance indicators. The real data should be compared with the budgeted metrics regularly by monitoring the key performance indicators daily to see that the metric is achieved equal or beyond it. It is required for the information systems to monitor and keep track of the performance of the entire workforce and link remuneration to it. Benchmarks are formed when something beyond a target is achieved. Right from the stages of exploration information technologies plays a vital role because if the information on the location of resources is not known, personnel in exploration will drill anywhere and may not find any resources causing investments with no returns at. It will cost manpower, equipment, and every other resource and still no returns, which is a fact with many mining investors.

Fuel management systems are an example of information system efficiency in open-pit mining. It starts from the point of receipt of fuel to the point of consumption by individual equipment or fleets. Data are completely available online for a huge network of fuel consumption in the mine site, for processing plant equipment, or running fleet. The data for the fuel used are integrated into the systems and available in real time as it is being used. Information on the fuel consumption, the time at which the fuel is required by a vehicle, calculation of the distance of the closest fuel tank, which vehicle requires when to fill the fuel, etc., are available within the system. It thus plays a huge role in productivity improvements. Digitalization helps in optimizing the time for fuelling, assist in data gathering for fuel consumption, determining the optimized level of fuel in different equipment, and assists in minimizing pilferages or thefts of fuel.

Information systems help in reducing the downtime and also improve the availability of fuel, the availability of utilization of fuel, fuel burn rate, etc., so that it has an impact on productivity improvements. In terms of fuel consumption, which is a major cost in mining, it is possible to make it very efficient using information technologies. Some systems use digital tags fixed on each equipment such that it is possible to know for each equipment the amount of fuel filled and at what time assisting in understanding the fuel consumption pattern of each equipment. This also eliminates the problem of accounting for fuel in which many organizations are also susceptible to theft.

Information technologies assist in **inventory management** by getting information on stock replenishments, minimum/maximum levels, and lead times, mode of shipment, transportation timeline, and current supplier. It assists the inventory controllers to determine what needs to be ordered, when and how much is the economic order quantities and helps in productivity in terms of making sure the materials are available when they are needed and the plant is not impacted on their production due to lack of materials. The logistics management system helps in knowing the online status of materials. It is separated from the inventory management systems, but the data are integrated into it. Inventory management starts from the planning stage while the logistics management system is giving a detailed insight into which part of the logistics chain, it is currently at. Once the material is ordered and it is on its way, it has a logistics chain to follow and this system gives the exact details of where the material is from the point that it is being ready at the supplier's point. The **warehouse management system** keeps track of stock in hand, coming stocks, stock in transit, the available demands (requisition raised by end-users), the available potential supplies in transit, and the stocks are under cycle counts.

The **drilling software Thunderbird** provides performance data, which includes the rate of penetration (ROP) for each meter of drilling and estimation of optimum drill parameters for a higher drilling rate. In drilling, there is a slight difference between the rate of penetration and productivity. There is an activity in between which is in the process but not productive but ROP gives in real-time the data on exactly when it is drilling. Thunderbird gives all drill parameters, most of them are recorded but all are in real-time, Thunderbird high precision GPS allows uploading of designed patterns directly to drills and the operator can take drill in position and drill. The human interface can be reduced, which makes it much safer and efficient. An **automatic drill depth indicator** is a software that gives continuous feedback to the drilling personnel on a mobile device about the depth at which drilling needs to be done because drilling more or drilling less than required, can increase costs or decrease productivity, respectively. It also assists in accountability as it can show which drill operator is not doing it right.

Data analysis and interpretation is generally a time taking process and requires specific skills and information technologies today can do this too. For example, there is a software called **MineVision** from Caterpillar that provides exact information's regarding improper transmission changes by an operator, or any kind of machine abuse, which will help the management to identify training requirements, machine status, etc., on a real-time basis. A **collision awareness system** is a GPS-based system that gives the position of each equipment to the mine personnel. It assists in information related to the direction of other equipment, alerts, and signals the personnel if any equipment is too close to it and helps in aversion of risks of collision in the open-pit.

There is software available for different areas in mining but some mining companies either do not want to have the cost of implementing it or training personnel for it.

Information technologies assist in understanding employee behavior by giving information on load times and cycle times of equipment to make appropriate decisions on manpower and equipment deployment. For example, leakage at the pipeline of the tailing dam is reported. In this case, under normal conditions, the personnel from the maintenance department needs to go and inspect it and then come back to take the tools and equipment necessary to repair it and then go back to the spot to repair the leakage. Implementation of information technologies and operational technologies like cameras can be extremely useful in such scenarios where the information can be received in terms of text and images or videos from the data collected through the cameras. The maintenance personnel can eliminate wastage of time by going to and from to bring the required tools and help to repair the pipeline and it reduces a lot of downtimes.

There are mobile apps available for recording absenteeism by allowing personnel to log into the system and inform of the time they are to arrive at the mine site or of leave they are to take which helps in planning for manpower resources in the mine site before the arrival of the mine personnel assisting in increasing the efficiency of time.

LSOPM operations require the common email system, communication system, Skype, and database servers that are common in any industry. Another general information system is the enterprise resource planning (ERP) system, which is a centralized system used for the internal purpose of the mining organization. LSOPM operations require specialized systems for resource modeling and geographical modeling which helps in mine planning for a better understanding of the ore body. One of the most important software is the expert system that is successfully used in process plants that produce multiple scenarios as solutions for a problem. The expert systems assist in decision-making and productivity improvements. The fleet management system in an LSOPM operation is considered as another critical requirement for a mining operation. It helps in monitoring the activity of the equipment in the pit and keeps track of its movement and performance. A list of information technologies specifically required for various areas in LSOPM operations is given in Table 2.6 in the appendix section at the end of the chapter.

2.5.3 Future Application Areas for LSOPM Operations

The advancement in mining technologies can be described in three stages: industrial engineering, remote-controlled mining, and automated and self-deploying systems (Chaykowski, 2001). The industrial engineering stage had advanced in process time and quality improvements, which was followed by the remote-controlled mining stage that introduced the computer-controlled mining systems and formed the basis of the third stage. The third stage of automated and self-deploying systems is the modern era of equipment and machinery performing the tasks without any human intervention. Automated systems require better teamwork, more specific and broad job descriptions, fewer employees at the operator level, and complete information sharing throughout the operations (Chaykowski, 2001).

Case Study Outcomes

The information systems in companies are overloaded with data—every time that someone puts a sensor in the next piece of equipment it sends a signal in terms of temperature, pressures and all of these are constantly measured second by second, sending that information to the WIFI, which sends it to the server, the server where it is stored and then someone gets to read it as the data from that sensor. That information where the sensor might be operating at 60 degrees, the revolutions are 100 per second or minute, the vibrations are certain numbers but it operates for 5 years, exactly with that same parameters, every second sending data and that data are sent every time. Currently, edge computing or higher computing devices are trying to get closer from information technology to operational technology to do the work. The computing power on the sensor needs to be programmed in a manner that when the revolutions go higher than certain level or lower than certain level then it sends the signals, or only when the temperature increases to a certain threshold value, send the signals or only when vibration monitoring senses something above the threshold value, it sends the signal. So that sensor may be in operation for five years and never send a signal, only sending signals to tell that it is still operating and when there is a problem it sends the signal.

There are certain areas where there is a scope of improvement for information technologies in LSOPM operations. The most common inhibition regarding information technology is that many mining companies do not efficiently use the systems implemented. Apart from that, it is also evident from various studies that the development of technologies in other industries is not yet realized or used in the mining industry, like the following:

- Augmented reality (AR)—Mining companies are new to the concept of AR and Virtual Reality (VR). AR and VR can assist in visualizing the entire mining area in 3D by exploring and analyzing the information from various integrated software in the company. AR can be used to have a 3D view of the ore body so that the exact locations of the ore can always be known to help in production activities. It could further assist in real-time block modeling based on drilling data provided to the database as the process takes place without the wastage in time for sampling or waiting to know the results of the sampling affecting decision-making processes and productivity. This can help in planning all resources, improving productivity, and optimizing costs keeping in mind the safety factor.
- Technologies like RFID (Radio Frequency Identification) and Barcoding, which is quite commonly used in the retail industry, is yet to be embraced by the mining industry in supply chain management and other areas. RFID is currently used in the SmartTag system, which is a system used in LSOPM operations for tracking the movement of ore from the mine site to the mill.
- Information systems to compare ore reserves, grade models, and actual mining data are also something that is not in use currently in the LSOPM area.
- Information technologies that can reveal the equipment condition in real time is an area that can help improve equipment efficiency further which is also not yet used in the LSOPM area.
- A reconciliation software that compares ore reserves, the grade models, the actual mining, and the mill results, all together in one seamless database, would be a useful tool that is currently unavailable in the market.
- Software for tracking of action items, for example, when there is a contract in it, the system tracks commitments, obligations, renewals, payments, etc. It requires training to understand and use the software effectively.

2.6 Conclusion

The use of information technologies in mining has been increasing over the years across all areas like it has advanced from nonintegrated information systems in supply chain management to integrated enterprise resource planning (ERP) and customer relationship management (CRM) managing the entire supply chain. The mining industry is changing, from robotic trucks to self-driving trucks, automation is considered as an enabler in boosting efficiencies and helping in keeping up with the safety standards. Robotic machines for mining can be operated from a distance remotely (Baggaley, 2017). They do not have to know what happened, instead, they now depend more on what happens now and what is most likely to happen. Countries that are ready for digitization have succeeded in the growth of new industries as well as the very swift development of local traditional sectors.

2.7 Summary

Information technology is considered a critical technology in mining to maximize productivity and minimize cost. Mining equipment is being automated increasingly and the mine operations area is becoming more and more integrated by communication and data networks to get more controlling capabilities in the field environment. The use of advanced technology also increased labor productivity and decreased the number of laborers required thereby decreasing the cost of labor. However, there is a huge potential in innovation and application of advanced information technologies in the LSOPM scenario and mining companies also need to harness the existing high-end technologies to achieve operational excellence.

Appendix: Additional Information on Information Technologies in LSOPM Operations

See Tables 2.3, 2.4, 2.5 and 2.6

Table 2.3 Role of information technologies in improving productivity

#	Description
1	Real-time information increases performance
2	Provides information to improve productivity
3	Assists in decision-making
4	Optimizes time

(continued)

Table 2.3 (continued)

#	Description
5	Processing data are better than manual systems
6	Better equipment performance
7	Cost optimization
8	Assists in increasing production
9	Communication in the field is easier and faster
10	Manpower skills can be improved
11	Assists in measuring the overall performance
12	Assists in minimizing risk
13	Assists in data analysis
14	Controls the mining process
15	Assists in understanding the ore body
16	Assists in managing human resources
17	Maximizes value of operation
18	Implementing the right IT and knowing its right use

Table 2.4 Role of information technologies in decision-making

#	Description
1	Helps in making decisions faster
2	Helps in providing information faster
3	Information is readily available in the systems
4	Helps in analyzing data
5	Real-time monitoring
6	Improves productivity
7	Manpower efficiency
8	Better decision-making processes
9	Helps in understanding equipment performance
10	Helps in setting benchmarks and goals
11	Helps in gathering field data
12	Helps in maximizing performance
13	Less dependence on manpower resources
14	Tasks are time consuming without information technologies
15	Helps in understanding the deposits better
16	Provides trending and historical data
17	Time efficiency

(continued)

Table 2.4 (continued)

#	Description
18	Better informed decisions on safety
19	Information technology is a tool not a strategy
20	Information technology not used by many mining companies efficiently
21	Helps in improving performance decisions
22	Must be used efficiently
23	Maximizes production
24	Crucial due to large area of operations
25	Mining operations rely heavily on information technology
26	Integration of operational technology and information technology
27	Information technology does not help in making better decisions

Table 2.5 Information technologies in LSOPM: current information technologies

#	Description
1	Fleet management systems
2	Enterprise resource planning systems
3	Resource modeling and mine planning software
4	Software for geological modeling and mine design
5	Communication systems
6	Email systems and skype
7	Database systems
8	Expert systems for mill
9	Information systems for drill and blasting
10	Software for statistical analysis
11	Geological software for monitoring blast movement
12	Plant process control and performance measurement systems
13	Geographical information systems
14	Survey software
15	Collision awareness system
16	Excel and sharepoint
17	Software for blast design
18	Fuel management systems
19	Use of drones
20	Operational technology and information technology integration
21	Customized software for risk assessment
22	Automatic drill depth indicator

(continued)

Table 2.5 (continued)

#	Description
23	Maintenance planning software
24	Dundas business intelligence software
25	Software for integrating drilling and survey data
26	Software for cyber risk
27	Equipment onboard computerized management systems
28	Application software for diamond drilling
29	Financial modeling software

Table 2.6 Specific information technologies required in LSOPM operations

#	Type of information technology	Description	Examples
1	Geological software	Analyze, model, process, and interpret data to visualize the geological structures for accurate Mine Planning	Leapfrog, Datamine, GEOVIA Surpac, Maptek Vulcan, Micromine
2	Survey software	The application used to analyze survey points to assist in design, model and interpret points on the ground relative to the other for survey purposes	Micromine, MineMap, Mine Survey Office (MSO)
3	Information systems for drilling and blasting	Methods to store, manage, document, and retrieve information related to drilling and blasting. Automatic drill depth indicator	BIMS (Blast Information Management system), Datamine, Thunderbird Mining Systems
4	Software for blast design and modeling	Application to design, charge initiate, analyze, and monitor blasts movement	SHOTPlus from Orica Limited, BlastLogic, BLASTMAP, JKSimBlast
5	Fleet management systems with collision awareness system and fuel management systems	To manage and improve fleet productivity and optimize costs; uploaded to interface with the equipment's dashboard and central database to track and monitor operator and equipment performance, collision avoidance ensure safety	Wenco Mining Systems, Modular Mining system, MineLink Fleet Management Systems (FMS)

(continued)

Table 2.6 (continued)

#	Type of information technology	Description	Examples
6	Plant process control and measurement systems	Systems to enhance, operate, measure, and monitor plant operations, control valves, process flows, flow volumes, temperature, and pressure gauges to optimize and improve plant performance and productivity	Distributed Control Systems (DCS), Supervisory Control And Data Acquisition (SCADA), EcoStruxure, SpectraMagic
7	Maintenance planning software	Application used to plan, organize, schedule, track, analyze, and report maintenance activities to enhance equipment life, improve productivity and lower costs	ERP Systems, Upkeep, Computer Aided Maintenance Management System (CAMMS), Wenco, eMaint CMMS
8	Integration of operational technology to information technology	Integrating cameras and sensors to software systems to get the required data	ERP Systems
9	Use of drones	Used in surveillance, collecting data, military, commercial, scientific, recreational, agriculture, photography, product deliveries, etc.	Unmanned Aerial Vehicle (UAV) Unmanned Combat Aerial Vehicle (UCAV)
10	Financial modeling software	Application to prevent error in financial analysis and provide visual graphs, charts to record, interpret, analyze and monitor financial matrix, Key Performance Indicators (KPI's)	MS Office—Excel, MapleSoft, Synario, Cognos,
11	Geographical information systems	Application for capturing, store, manage, and display data, coordinates, positions on the earth surface relative to the other point	ArcGis, MapInfo, GeoMedia, Manifold, MapViewer
12	Business intelligence software	Application to retrieve, analyze, transform, report, and present data for decision making	SAP, Oracle, Cognos, Qlik
13	Common software like Microsoft excel and Sharepoint	Application for users to input, analyze, interpret, report and store, manage, and share data	MS-Excel, MS-SharePoint

References

Adams, M. (2017). Big data and individual privacy in the age of Internet of Things (IoT). *Technology Innovation Management Review, 7*(4), 12–24.

Airnei, D., & Berta, D.-A. (2012). Semantic business intelligence—A new generation of business intelligence. *Informatica Economică, 16*(2), 72–80.

Australian Mining. (2016, August 17). *The top ten trends for mining in 2016: Australian Mining*. Retrieved from Australian Mining Web site: www.australianmining.com.au.

Baggaley, K. (2017, December 21). *Robots are replacing humans in the world's mines. Here's why: NBC Universal*. Retrieved from NBC Universal Web site: https://www.nbcnews.com.

Baltzan, P., & Welsh, C. (2015). Decision making and business processes. In P. Baltzan & C. Welsh (Eds.), *Business driven information systems* (pp. 30–62). McGraw Hill Education.

Barrick Gold Corporation. (2018, May 15). *High tech mining in the gold belt: Barrick gold corporation*. Retrieved from Barrick Gold Corporation: https://www.barrick.com.

Bazzazi, A. G., Osanloo, M., & Behrooz, K. (2009). Optimal open pit mining equipment selection using fuzzy mutiple attribute decision making approach. *Archives of Mining Sciences, 54*(2), 301–320. Retrieved from https://mining.archives.pl/.

Burt, R. (1991, May). The international diffusion of technology in the early modern period: The case of the British non-ferrous mining industry. *The Economic History Review, 44*(2), 249–271. doi:https://doi.org/10.2307/2598296.

Business First Magazine. (2017). *Business operational excellence: Business First Magazine*. Retrieved from Business First Magazine Web site: https://www.businessfirstmagazine.com.au.

Campbell, B. L. (2007). Exploration, production call for new technologies. *Canadian Mining Journal*, 9–11.

Carter, R. A. (2006, May). Hybrid system solves GPS problems at morenci. *Engineering & Mining Journal*, 46–49.

Chaffey, D., & White, G. (2012). *Business information management*. Edinburgh: Pearson Education Limited.

Chanda, K. E., & Steven, G. (2010). A comparative study of truck cycle time prediction methods in open-pit mining. *Engineering, Construction and Architectural Management, 17*(5), 446–460. https://doi.org/DOI10.1108/09699981011074556

Chaykowski, R. P. (2001). Re-inventing production systems and industrial relations: Technology-driven transformation in the Canadian metal-mining industry. *Journal of Labour Research*, 591–613.

Dassault Systems GEOVIA. (2015). *Home: Dassault systems GEOVIA*. From Dassault systems GEOVIA: www.geovia.com.

Datamine. (2015). *Home: Datamine*. From Datamine Web site: www.dataminesoftware.com.

Dinh, T. L., & Fillion, G. (2007). Acquiring domain knowledge of information systems: The information system upon information systems approach. *Allied Academies International Conference* (pp. 19–24). Reno: The Academy of Information and Management Sciences.

Diss, K. (2015, October 18). *Driverless trucks move all iron ore at Rio Tinto's Pilbara mines, in world first: ABC*. Retrieved from ABC Web site: https://www.abc.net.au.

Dold, B. (2008, September 27). Sustainability in metal mining: from exploration, over processing to mine waste management. *Reviews in Environmental Science and Bio/Technology*, 275–285. doi:https://doi.org/10.1007/s11157-008-9142-y

Donnelly, H. J., Gibson, L. J., & Ivancevich, M. J. (1978). *Fundamentals of management: Functions, behavior, models*. Dallas: Business Publications Inc.

Durugbo, C., Tiwari, A., & Alcock, J. R. (2011). A review of information flow diagrammatic models for product–service systems. *International Journal of Advanced Manufacturing Technology*, 1193–1208.

E&MJ. (2009). New tools put data to work. *Engineering & Mining Journal*, 68–71.

E&MJ. (2012a). Australian technology: technology drives improved productivity. *Engineering & Mining Journal*, 102–115.

E&MJ. (2012b, March). Chilean service companies become world class. *Engineering & Mining Journal*, 68–80.

Elko Daily. (2018, June 5). *Barrick launches open pit autonomous haul trucks in Nevada: Elko Daily*. Retrieved from Elko Daily Free Press Web site: https://elkodaily.com.

EY. (2013). The future of decision making. *5 Insights for Executives*, 1–8.

EY. (2017). *The Digital mine: EY*. Retrieved from EY Website: www.ey.com.

Fiscor, S. (2010, May). Technology allows mine planners to consider more detail. *Engineering and Mining Journal, 211*(4), 40–42. Retrieved from https://search.proquest.com/business/docview/357029492?accountid=150425#.

Fiscor, S. (2015). Haulage and loading offers educational programs for open-pit mining. *Engineering & Mining Journal*, 60–70.

Fiscor, S. (2016a, July). Advances with large rotary Blasthole drills. *Engineering & Mining Journal*, 26–29.

Fiscor, S., & Casteel, K. (2008). Australia exports mining and processing technology. *Engineering and Mining Journal*, 62–71.

Furterer, S. (2009, April). Secure operational excellence. *Industrial Engineer, 41*(4), 43–47. Retrieved from https://search.proquest.com/business/docview/231493455/A334DBFAF0474288PQ/14?accountid=150425.

Giddaluru, M. P., Gao, J., Bhatti, R., & Shah, S. (2013). Requirements analysis in the implementation of integrated PLM, ERP and CAD Systems. *11th International Conference on Manufacturing Research* (pp. 45–50). Cranfield University.

Hanson, D. (2016, June 1). *Problem solving: A look at analyzing mining data to improve performance*. Retrieved from Canadian Mining Journal Web site: www.canadianminingjournal.com.

Hartman, L. H., & Mutmansky, M. J. (2002). Introduction to mining. In *Introductory Mining Engineering* (2 ed., pp. 1–21). New Jersey: John Wiley & Sons, Inc.

Haslam, C. (2007). *Business analysis and decision making*. London: The University of London Press.

Hicks, G. H., & Gulliett, R. C. (1981). *Management*. New York: McGraw Hill Inc.

Hitt, M., Black, S., & Porter, W. L. (2014). Individual and Group Decision Making. In M. Hitt, S. Black, & W. L. Porter (Eds.), *Management* (3rd ed., pp. 325–350). Edinburgh Gate: Pearson Education Limited.

Hiyate, A. (2017, February). Pushing boundaries: A snapshot of mining R&D activity in Ontario. *Canadian Mining Journal*, 20–25.

Jayashankar, R. (1989). *Decision support systems*. New Delhi: Tata McGraw-Hill Publishing Company Limited.

Jordaan, J. T., & Hendricks, C. (2009). The challenge of technology adoption and utilisation in the mining industry—A focus on open pit mining technologies. *Base Metals Conference 2009* (pp. 69–82). The Southern African Institute of Mining and Metallurgy.

Kadiyala, R., & Kleiner, B. H. (2005). New developments concerning business information systems. *Management Research News*, 164–170.

Kramer, J. W., Jenkins, B., & Katz, S. R. (2007). *The role of the information and communications technology sector in expanding economic opportunity*. Harvard University. Retrieved from https://www.hks.harvard.edu/m-rcbg/CSRI/publications/report_22_EO%20ICT%20Final.pdf.

Lee, E., Oh, J.-Y., & Pines, E. (2008). Practical managerial decision making tools: operations research. *The Journal of Applied Business and Economics*, 11–18.

Lee, H., & Whitley, A. E. (2001). Time and information technology: Temporal impacts on individuals, organizations, and society. *The Information Society*, 235-240. doi:https://doi.org/10.1080/01972240290075084.

Lewis, M. W., & Steinberg, L. (2001). Maintence of mobile mine equipment in the information age. *Journal of Quality in Maintenance Engineering*, 264–274.

Lovrencic, S., Sekovanic, V., & Antonio, J. (2017). Knowledge management in the wealth of information technology. *26th international scientific conference on economic and social development* (pp. 499–507). Zagreb: 26th International Scientific Conference on Economic and Social Development.

Maptek Pty Ltd. (2018). *Introducing Vulcan 10.1*. Retrieved from Maptek Website https://www.maptek.com/insight/whats_new_in_vulcan10.1.html.

Marcus, J., & Dam, V. N. (2015). Decision-making. In J. Marcus, & V. N. Dam, *Organisation & management: An international approach* (3rd ed., pp. 345–384). Houten: Noordhoff Uitgevers bv.

McCrohan, K. F., & Harvey, J. W. (2008, November 1). Current issues in the role of technology in emerging market strategy. *Journal of Transnational Management, 13*(1), pp. 3–22. doi:https://doi.org/10.1080/15475770802059388.

McNab, K., & Garcia-Vasquez, M. (2011). *Autonomous and remote operation technologies in Australian mining*. Brisbane: The University of Queensland.

McRoberts, P. (2016, September). World mining equipment: Operating strategies-counting on the connected mine. *Engineering & Mining Journal*, 158–164.

Mining Global. (2015, August 12). *GoldCorp—Operational excellence done right: Mining global*. Retrieved from Mining Global Web site: www.miningglobal.com

Naicker, V., & Jairam-Owthar, D. (2017). The linkage of information quality to an executive decision support framework for the financial service sector of a developing economy. *South African Journal of Information Management*.

Nowduri, S. (2010). Management information systems and business decision making: review, analysis, and recommendations. *Journal of Management and Marketing Research*.

Pan, J., Oates, C. J., Ihlenfeld, C., Plant, J. A., & Voulvoulis, N. (2010). Screening and prioritisation of chemical risks from metal mining operations, identifying exposure media of concern. *Environ Monit Assess*, 555–571.

Peterson, D. J., LaTourrette, T., & Bartis, J. T. (2001a, February). New forces! Industry views of critical technologies. *Engineering & Mining Journal*, 46–48 & 59.

Procom. (2012, December 7). *Information technology helps spur innovation in Canadian mining industry*. Retrieved from Procom Web site: https://www.procom.ca.

PwC. (2015). *Mining excellence in Africa: PricewaterhouseCoopers*. Retrieved from PricewaterhouseCoopers Web site: www.pwc.com.

PwC. (2017). *We need to talk about the future of mining: PricewaterhouseCoopers*. Retrieved from PricewaterhouseCoopers Web site: www.pwc.com.

Ramrathan, D., & Sibanda, M. (2017). The impact of information technology on intuition in organisations: A phenomenological approach. *The Journal of Developing Areas, 51*(1), 207–221.

Rohleder, C., Davis, S., & Gunther, H. (2005). *Issues in Information Systems, 6*(2), pp. 345–351. Retrieved from https://iacis.org/iis/2005/Rohleder_Davis_Gunther.pdf.

Roman, P. A. (1999). *Joint enhancement of mine operations and maintenance: A structured analysis approach*. Queen's University (Canada). Ann Arbor: ProQuest, UMI Dissertations Publishing. Retrieved from https://search.proquest.com/business/docview/304542219?accountid=150425.

Tank, M. D. (2015). Enable Better and Timelier Decision-Making Using Real-Time Business Intelligence System. *I.J. Information Engineering and Electronic Business*, 43–48. Retrieved from https://www.mecs-press.org/ijieeb/ijieeb-v7-n1/IJIEEB-V7-N1-6.pdf.

The service sector. (2012, May). *Engineering and Mining Journal, 213*(5), 132–136,138–145. Retrieved from https://search.proquest.com/business/docview/1021202757?accountid=150425.

Thompson, I. (2015, December 1). *Harnessing information technology to drive productivity gains: Canadian Mining Journal*. Retrieved from Canadian Mining Journal Web site: www.canadianminingjournal.com.

Thunderbird Mining Systems. (n.d.). *Drill Information Systems: Thunderbird Mining Systems*. Retrieved 2017 from Thunderbird Mining Systems Web site: www.tbirdpac.com.

Uden, L., & He, W. (2016). How the internet of things can help knowledge management: A case study from the automotive domain. *Journal of Knowledge Management, 21*(1), 57–70.

Warr, A. (1990). *Bridging the gap—Implementing Information Systems (IS) strategies*. Bedford: Cranfield University.

Wenco. (2017). *Industry news: Wenco international mining systems Ltd.* Retrieved from Wenco International Mining Systems Ltd. Web site: https://www.wencomine.com

Whyte, D. H., Geraghty, R., Pujol, F., & Sellschop, R. (2015, November). *How digital innovation can improve mining productivity: McKinsey & Company*. Retrieved from McKinsey & Company Web site: www.mckinsey.com.

Whyte, G., & Bytheway, A. (1995). *Factors affecting information systems success*. Bedford: Cranfield University.

Yamakawa, S. (1997). *The development of a framework for inter-disciplinary building design working, and the application of intelligent knowledge-based system techniques*. Cranfield University.

Yudelman, D. (2006a). Advanced mine technology: Some examples. *Canadian Mining Journal*, 17–22.

Yudelman, D. (2006b). Mining technology: Scratching the surface. *Canadian Mining Journal*, 5.

Chapter 3
Organizational Efficiencies and LSOPM Business

Abstract Organizational resources need to match up with the organizational requirements and needs to be used efficiently. Organizational efficiency can be described as obtaining maximum output by using minimum resources. Productivity and cost optimization are two factors that are relevant and have been elaborately explained in this chapter. This chapter also deals with the concepts of productivity and cost optimization and specifically showcases the results of the primary research in terms of the interrelationship between costs and productivity. Further, this chapter also deals with the relevance of equipment and its efficiency in mining by representing it in the form of maintenance, equipment availability, and utilization.

Keywords Productivity and cost optimization · Mining equipment efficiency · Equipment availability and utilization

3.1 Introduction

Operational excellence is not a separate concept, showing how to run the business to achieve the success envisioned by the people in the business (Chevron Corporation, 2010). Operational excellence results can be continually improved by using a standard approach to systematically identify and close the performance gap (Chevron Corporation, 2010). Athenian Brewery SA, one of the largest companies in Greece, is a role model for other organizations in their region and industry, pioneering initiatives and implementing them to achieve operational excellence (Vrellas & Tsiotras, 2014). Athenian Brewery used the management philosophy of total quality management and total productive management program to drive continuous improvement and focuses on product quality and safety, environmental protection, and employee health and safety (Vrellas & Tsiotras, 2014).

Organizations that use modern management philosophies like lean production and six sigma also use models that are divided into many phases, where typically the last phase is the operational excellence phase (Mast et al., 2013). The last phase is a phase that continues after implementation and integration and can be understood as the monitoring and control phase where improvements are continued as new changes occur in the environment due to recent developments (Mast et al., 2013). Operational

H. Qudrat-Ullah et al., *Operational Sustainability in the Mining Industry*,
Asset Analytics, https://doi.org/10.1007/978-981-15-9027-6_3

excellence can be achieved if organizations establish, effectively communicate, and periodically assess the requirements (Bigelow, 2002). Management concepts like lean engineering, JIT, and six sigma can help bring in enhancements in operational efficiency, but they cannot eliminate the day-to-day operational costs that arise from various factors contributed to human error. Management in an organization must be aware of all systems and processes to be able to recognize the strengths and weaknesses of the organization. Management in organizations must be committed to quality, continuous improvement, and total compliance (Bigelow, 2002).

Operational excellence can be achieved by the philosophy of continuous improvement, a strong focus on the customer, and a continuous need for the employees to be able to identify and correct problems (Ion et al., 2015). The goal of operational excellence is not however continuous improvement initiatives that are employed to reduce costs and increase productivity; rather it is business growth that will allow management to deeply integrate into innovation and guide product development, as well as supply chain, manufacturing, and technical capabilities (Duggan, 2013). Instead of using management philosophies and management intervening in setting goals and objectives, businesses should set a destination of operational excellence and reach it by applying a specific set of technical guidelines and principles (Duggan, 2013).

Operational excellence is not just about using tools like six sigma, kaizen, and other important continuous improvement techniques (Rusev & Salonitis, 2016). It includes lean principles with organizational culture and management at a strategic level and has been defined as an enterprise-wide practice where every employee can see the flow of value to the customer and fix that flow before it breaks down. It can be understood from the above that management philosophies like continuous improvement, lean, six sigma, and JIT may be implemented and used in organizations but it is important first to understand the productivity and cost concepts for an organization to sustain operations (Rusev & Salonitis, 2016). The chapter will consider two important factors to understand operational excellence—the cost factor and the productivity factor and added to it is a third factor, the equipment factor. This section is further divided into four subsections:

- Organizational efficiencies
- Organizational efficiencies in mining
- Equipment in mining
- Organizational efficiencies in LSOPM operations

3.2 Organizational Efficiencies

Organizational efficiency relates to operational excellence that is defined as a philosophy at the operational level with a stronghold on competitiveness at the strategic level and can be expressed in terms of three important areas: focus in the process, focus on customer, and world-class performance (Ion et al., 2015). Organizational efficiencies relate to efficiencies in terms of production, costs, employee, performance, process, and time. Each of these efficiencies has been explained further in this section to distinguish and understand how each contributes to an organization's long-term growth and sustainability.

3.2.1 Organizational Productivity

Conventionally, productivity is measured by two important factors—labor productivity (in terms of output per capita/per person employed/per employed hour) and total factor productivity, which is the portion of output not explained by the number of inputs used in production. The five drivers of productivity as per the UK government regulations are an investment, innovation, skills, enterprise, and competition (Mayhew & Neely, 2006). Profitability is a measure to show an organization's performance and productivity improvements are important to enhance profitability. The impact on profitability can be seen by the management by closely monitoring productivity changes and putting it in monetary terms as is done with profitability (Cosgrove, 1986).

Productivity growth comes from static efficiency, which is using the current factors of production as effectively as possible or dynamic efficiency, which is about physical investment in new machinery, technical progress, knowledge (R&D), and human beings in terms of professional development and human capital (Mayhew & Neely, 2006). Production management deals with the way the products and services will be made or provided and the managerial activities are required to achieve it (Hicks & Gulliett, 1981).

The focus of production management is on manufacturing technology, the flow of materials, and the production of goods. The production function in any company focuses on the activity of producing goods that deal with the design, implementation, operation, and control of men, materials, equipment, money, and information to achieve specific production objectives (Donnelly et al., 1978). An organization's operations and production management department find information vital to run its business processes and the performance of an organization depends largely on improvements made by reviewing and making changes accordingly to make the processes more effective and efficient (Chaffey & White, 2012).

Many organizations try to follow a given method but fail to understand that this cannot ensure good performance because performance is not absolute but relative in a competitive environment (Singh, 2012). Many times, improvements in operations through better quality, lower cost, swifter throughput time, and high-quality asset management may still not ensure better performance because if the organization cannot attain a competitive advantage, then the performance is understood to be suffering (Singh, 2012). To ensure the optimization of resources in organizations, management teams should pay attention to the use of managerial science, statistics, and quantitative approaches for planning, forecasting, cost analysis, and evaluation (Itanyi et al., 2012).

Productivity tools such as behavioral methods and quantitative techniques are important and should be applied to problem-solving in organizations (Itanyi et al., 2012). A new term often used to measure the performance of an organization is business excellence, a measure used to have excellence in strategies, business practices, and performance results concerning stakeholders (Mann et al., 2011). The EFQM Excellence Model and the Baldrige Criteria for Performance Excellence are two types of awards that form as assessment tools to measure the level of distribution in business excellence within an organization (Mann et al., 2011). Performance excellence is the result of the intelligent use of organizational resources and capabilities, which are the main reasons determining competitive advantage (Asif & Searcy, 2014).

Productivity measurements vary from company to company, industry to industry, and country to country, which demonstrates a vast difference in productivity even though many factors remain the same. One big reason for this difference is the use of different technologies and the additional complexity of measuring intangible assets like organizational, reputational, and managerial capital (Griffith et al., 2006). It is important to use the correct measures to know the status of the operations about productivity otherwise it will prove as a barrier to improve resource productivity (Armistead, 1991).

People expect technology can cause improvements in productivity and that it can enable better control of capacity in line with demand and thereby meet the prerequisites of customer service and resource productivity targets (Armistead, 1991). Productivity, competitive advantage, and improvements are all interrelated because international competition and economic growth rely on resources being used efficiently (Britney et al., 1986). Management should not rely on unsupported data evidence regarding productivity because to make accurate changes in productivity factors, data, and information about it should be readily available (Cosgrove, 1986).

The right combination of people, process, and technology coming together to enhance the productivity and value of any business operation, at the same time bringing down the costs to a planned level help in achieving operational efficiency. The remaining resources can then be redirected to initiatives that are of high value to bring in additional profitability to the business. Intangible resources like an organization's reputation can be a valuable resource for competitive advantage (Kanghwa, 2010).

Competitive advantages are the result of gaining a more favorable position on the market, and to achieve it organizations must continuously develop plans designed to ensure the long-term sustainability of competitive advantages, either by using the organization's resources or by using its capabilities (Vele, 2014). Employee involvement is the key to ensuring the long-term sustainability of competitive advantage as employees are the most important resource especially when it comes to strategies based on innovation and are responsible for the formulation and implementation of competitive strategies. Resources, which are (a) rare and decisive in satisfying customer needs, (b) responsible for a large percentage of profits, and (c) have a low depreciation value, are valuable (Vele, 2014). Using a reduced amount of resources conserves expenditure thereby reducing costs and minimizes wastes (Driussi & Jansz, 2005).

3.2.2 Organizational Costs

Cost analysis is particularly important for any organization and it is vital for understanding the feasibility of a product and the effort needed to produce it (Ball et al., 2011). Cost leadership can be achieved in an organization that considers three related components: (a) tools, like ABC and six sigma, which help companies to understand and manage costs; (b) metrics tied to compensation, like customer profitability and cost to serve; and (c) building and sustaining a culture that supports in attaining and sustaining cost leadership (Schiff, 2014). The costs, revenues, and profits are related to one another as shown in Fig. 3.1 (Hustrulid et al., 2013).

To attain cost leadership, the focus should be on efficiency, an example of which can be the simple outcome of a focus on controlling cost by the manager, employee productivity, and the economical use of an asset. Sustaining a cost leadership culture within a company can have excellent effects in the long term for the company (Baack & Boggs, 2008). One such example of this is TPS from Toyota that emphasizes employee participation by 100% leading to long-term growth, the best quality in the industry, and high-profit margins (Schiff, 2014).

For the cost leadership strategy to work, employees in an organization should have confidence, support, acumen, and they need to understand how the values impact their work and the tangible added value from the work (Schiff, 2014). To gain competitive advantage, by using a cost leadership strategy, the organization should use the following measures: (a) use its resources to lower its production cost by increasing the employees' productivity; (b) use more efficient suppliers, adapting production and distribution systems that are focused on lowering the operation costs; and (c) use production facilities that will lead to economies of scale. Productivity cost estimates in an organization should also include costs related to absenteeism, presentism, and unpaid work (Krol & Brouwer, 2014).

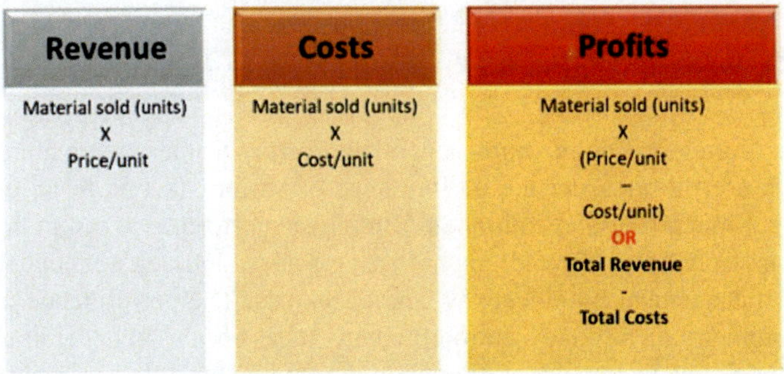

Fig. 3.1 Relationship between costs, revenues, and profits

3.2.3 Employee Efficiency

Education and training play an important role in an organization where employees are concerned. Employee efficiency requires measuring the value added by manpower resources based on business needs and life of mine. The fact that the cost associated with the learning curve for a new employee may be greater than retaining a good employee is overlooked by businesses many times (Vele, 2014). The learning curve of new employees deals with the length of time it takes for a new employee to be fully operational in his new role in terms of understanding the business, the new work environment, and its dynamics and understanding the improvements that can be made based on their abilities and their experience. This in turn can be called the hidden cost because retaining good employees would assist in a good return on investment.

3.2.4 Process Efficiency

Process efficiency is related to the amount of effort required to achieve business objectives. For processes to be efficient, the first and foremost thing is to understand the primary business goals a process needs to achieve. Apart from that, it is important to understand all the constraints the process must operate within and the strengths and weaknesses of the process in the current state (Valery & Jankovic, 2011). Process efficiency can be achieved by business process optimization. The stages of the business process optimization consist of about seven stages that are defining outcomes, defining the current state, identifying gaps, selecting test cases, developing/testing hypotheses, implementing new processes, evaluating performance.

3.2.5 Performance Efficiency

Performance efficiency or in more general performance measurement relates to the methods used to measure the performance of various factors in an organization including manpower and equipment. Employee performance can be measured using various techniques like the graphic rating scales, 360-degree feedback, self-evaluation, management by objectives, and checklists. The performance measurement of equipment can be done by how efficiently it has been used as per its available capacity considering the factors of its economic lifetime, breakdowns, maintenance, and maintenance techniques (Lewis & Steinberg, 2001).

3.2.6 *Time Efficiency*

Efficiency in time management relates to eliminating wait times or lag time between various processes. In management, time is related to money and every wait time between processes costs the organization. Effective time management systems using time management software can assist in providing information faster for employees to manage time and increase efficiency in time management. Apart from that, effective process management systems, systems in operations, and monitoring performance are a few steps that enhance efficiency related to time (Singh, 2012).

3.3 Organizational Efficiencies in Mining

The income of a mining project depends on the volume of the product and its price. A mining firm has three basic objectives as illustrated in Fig. 3.2 (McIlroy, 1999). The profitability of a company depends on the revenues, which should be much more than the costs incurred in terms of development cost, production cost, and costs associated with discovery and delineation of economic mineral deposits. The survival of a mining company depends on the economic mineral deposits that the company obtains. The growth of a mining company includes an increase in share price also known as market capitalization, number of mines, the amount of metal produced, and sales revenue (McIlroy, 1999).

3.3.1 *Productivity in Mining Operations*

The typical choice of a mining investor is to invest in a high grade easily accessible deposit because it is assumed that the extraction is easier and will be profitable due to the grade of the deposits and its quality (Topp et al., 2008). The requirement of

Fig. 3.2 Basic objectives of a mining firm

each mining company according to its geographic, geological, and resource availability is different and so the technique to improve productivity in each company is different (Britney et al., 1986). Focus on productivity should be for the long term to have sustainable improvements that will require a change in organizational culture and thinking processes. For long-term productivity, the companies should be ready to make changes in the form of (a) change in mine plans, (b) reassessment of mining methods, (c) make changes to equipment fleet and configuration, (d) reduce production, and (e) increase or reduce automation (Fazal, 2014).

It is important to understand that top-management structure and hence strategic decisions are the key to make changes for improving productivity, however, employees from all levels must be engaged to bring in operational excellence in a mining scenario (Topp et al., 2008). The main objective of the business remains the same, but changes can be made to the operational strategies of the business to bring forth improvements in productivity. Productivity is generally defined as the rate of production in a specific time interval. It is a measure of efficiency and performance of the system. There are many reasons for increasing or decreasing productivity and thus this concept is popularly known as multifactor productivity or MFP (Topp et al., 2008). Labor productivity, capital productivity, cost reduction initiatives are some of the factors that must be considered for improving productivity (Mitchell & Steen, 2017).

In large-scale open-pit mining, a recent methodology for process efficiency is the PIO or Process Integration and Optimization methodology, which deals with improving productivity by optimizing processes in a step-by-step manner throughout the complete mining process (Valery & Jankovic, 2011). The PIO methodology involves all processes from drilling, blasting, loading, hauling, dewatering, crushing, milling, and other such activities in the production process of the mining operations. It requires a thorough understanding of the current operating practices; stresses understanding the deposits in terms of rock or ore characterization and helps in deciding on realistic targets and benchmarks to achieve high operational efficiency (Valery & Jankovic, 2011).

One of the factors related to productivity is a capital investment that needs to be done smartly otherwise a poorly applied investment can result in decreasing productivity (Lala et al., 2016). Productivity in mining is not only related to just the efficiency in production but also enhanced by the accessibility and quality of the resources available in the deposit (Topp et al., 2008). Production efficiency is related to technological advancement and better management techniques (Lala et al., 2016). The decrease in production can increase productivity if appropriate measures to optimize costs, labor, and capital take place. It is important to understand what factors you consider when deciding on the trade-off. For example, a piece of old equipment may help in less capital investment but can increase maintenance costs considerably (Lala et al., 2016).

Mining companies need to focus on productivity because it has been declining since the 2000s (Mitchell & Steen, 2017). The decline in productivity depends on resource reserves and economies of scale. Mining companies should strive to achieve higher productivity with lower inputs while raising output leads to greater improvement in productivity (Mitchell & Steen, 2017). One of the measures to keep productivity decline in check is for the mining companies to focus continuously on throughput improvements and reduction in costs which will eventually help in achieving operational efficiency (Lala et al., 2016). However, another important measure for continuously monitoring is a digitization and the use of new and advanced technologies to have real-time data that help in various areas of mining like equipment maintenance and utilization, analyzing variables for statistical purposes in process plants, etc. While cost is an important factor when productivity is concerned, cost cutting does not ensure productivity, and to improve productivity, the investments must be such that the resources are used to the maximum (Dassault Systems GEOVIA, 2015).

3.3.2 Costs in Mining Operations

Profit in a mining operation over a period can be increased by the flow of high-grade material to the processing plant and this is where the cut-off grade is an important factor (Asad & Topal, 2011). Mining costs depend on various factors like ore processing techniques, equipment selection, drilling methods, blasting methods, and haulage. Ore processing technique involves choosing the option of processing the ore on-site, shipping the concentrated ore directly to the market, or choosing other off-site processing plants of the company (Miguel, 1996).

Grain size is an important factor in recovery because small and fine grains of minerals are difficult or impossible to recover in the processing plant (McIlroy, 1999). If the minerals are hard, then it can also result in being uneconomical because hardness will increase the grinding cost. It is never economically practical to extract all available ore from the mine. The cost factor plays an important role in mine recovery and processing recovery because it is overly expensive to mine and process the last remaining tonnes of ore from the mines. It is profitable to recover most of the mineral at less cost than to recover all of it at an extremely high cost (McIlroy, 1999). The supply process of minerals has five areas of costs as illustrated in Fig. 3.3 (McIlroy, 1999).

Many studies have two kinds of costs that are generally considered in mining, the operating cost and the capital cost (Miguel, 1996). The operating costs are further into direct, indirect, and general (overhead). The direct costs include the cost of labor, material, royalties, and development. Labor and material can be further categorized into labor—direct operating, operating supervision, direct maintenance, maintenance supervision, and payroll burden on the foregoing labor; materials—maintenance, repair materials, processing materials, raw materials, consumable energy (gas, diesel, oil, water, water, power, etc.) (Miguel, 1996).

The indirect operating costs are for labor, insurance, depreciation, interest, taxes, reclamation, travel, meetings, donations, office supplies, upkeep, utilities, and general mine development (Miguel, 1996). The labor cost in indirect cost includes administrative, safety, technical, service (clerical, accounting, general office), shop and repair

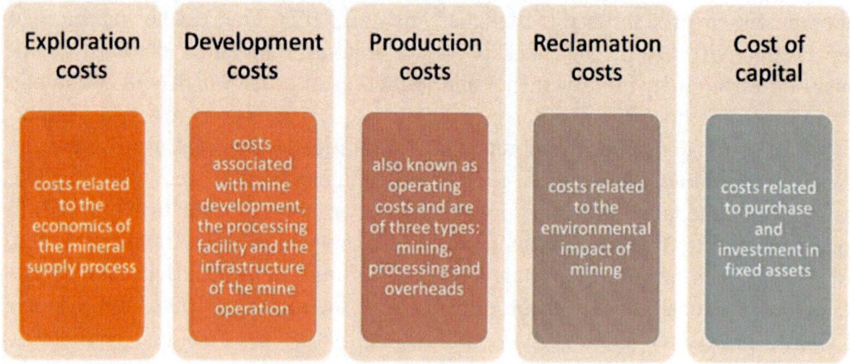

Fig. 3.3 Areas of costs in supply process of minerals

facilities, and payroll burden on the foregoing. The general expenses are usually for marketing and administrative areas (Miguel, 1996).

Cost components for mining operations can be distinguished during the mine planning stage in the form of fixed costs and variable costs (Steffen, 1997). The cost model should include total mining costs, which are calculated by adding up fixed costs, variable costs and all mining equipment capital costs discounted into the unit mining operating cost. In mining operations, material handling is of prime importance because it is one of the most significant factors for the operational cost (Burt et al., 2011).

The objective of any mining company is to increase its net present value or NPV of its cash flow, which can be achieved by optimizing long-term mine extracting sequences and the use of materials that have been mined (Goodfellow & Dimitrakopoulos, 2016). The planning process for an open-pit mine has the goal of finding the optimum annual schedules that will give the highest NPV while meeting various types of constraints related to production, blending, sequencing, and pit slope (Dagdelen, 2001). The process of scheduling starts with the assumption of the beginning capacities of production and the estimates for the related costs and the prices of the commodity (Dagdelen, 2001).

After the economic parameters are determined, the ultimate pit limits of the mine are analyzed to decide the portion of the deposit which can economically be mined. The design is further refined by dividing the deposit into nested pits wherein the smallest pit is with the highest value per ton of ore and the largest pit is with the lowest value per ton of ore (Dagdelen, 2001). Until a few decades, when high-grade reserves in a mining operation were adequate to supply the market needs, the philosophy was to remove the material in a particular order, such that the high-grade ore body is used until depletion and since the profits were always high this method was never questioned before (Johnson, 1968).

Mining companies sometimes have the limited know-how of concepts of cost cutting across the mining operation and often tend to have a deficiency in financial and technical understanding of the underlying issues about costs (Peterson et al.,

2001c). Capital costs in mining operations have been at an all-time high for the past few decades, the reason for which include: the desire of companies to rush into operations too fast, the obvious increase in costs of equipment and materials, issues with labor, and the dearth in people in the management who are effective, knowledgeable, and experienced (CMJ, 2012).

Maintenance costs can contribute from 40 to 80% to the overall operating costs in mining companies (E&MJ, 2012a). However, if the price of various metals is considered, then at the end of 2015, in comparison to the end of 2014, the price of iron ore went below by 24%; palladium went below 30%; copper went below 25%; zinc went below by 30%, and aluminum went below by 19%, with only gold recovering slightly from the lows of 2015. Many countries around the world spend millions of dollars in mining R&D and innovation, for example, Canada spends $550 million per year and Australia spends $2.8 billion (Hiyate, 2017).

The cost of the exploration and feasibility study for a large mining project can be $50 million or more, which is a cost incurred even when it is uncertain whether the deposit will be profitable or not in the future (Mikesell & Whitney, 2016). Thereby, reducing costs of exploration and production-related activities has become vital for mining companies planning for long-term sustainability and there is a growing need for technologies that can perform these activities (Hiyate, 2017). The cost can be projected using software packages used for simulation of the mining operations, which produces a pragmatic schedule of long-term production costs and costs can be distinguished as the cost for energy, maintenance, labor, and capital (Steffen, 1997).

Development of computer software packages has paved the way to a greater analytical capability for mining engineers and open-pit mining becomes most effective for the huge tonnage of material because of the high degree of mechanization that can be accomplished, which results in low unit operating costs making lower grade deposits more economic (Steffen, 1997). Mining engineers use computer software packages nowadays to perform the analysis of the ultimate pit limit and decide yearly mine plans and schedules. The mine plans produced with the help of the modern software packages are most of the time feasible and are implemented in mining operations, obtaining the highest returns possible on the capital invested in overlying waste blocks (Dagdelen, 2001).

Case Study Outcomes

When considering mining costs, for example, in terms of contract negotiation in supply chain management, contracts should be monitored and unwanted contracts must be reviewed and negotiated from time to time. If we take the example of a gold mining company at present in the case of activated carbon which is one of the key products used in gold mining, the market is showing that there are a scarcity and huge pricing and supply fluctuations in coconut shell-based carbon. Then the time is not right to go for a long-term contract at this stage, but rather a short-term contract is a better option, and keeping a close watch of the market is important. Long-term contracts are preferable when the markets are favorable (Buyer's market).

In mining, for example, if information technologies on the vendor's side are not aligned with the organization's supply chain department, this could end up in major time and cost issues. Contracts and negotiations for procurement of spare parts of equipment and the heavy equipment itself are handled by this department and this requires the personnel in the department to be aware of the situations in the mining area. They must be in constant contact with field personnel to be able to understand the right specifications to procure the right equipment, spare part, component, etc.

Along with optimization, one important method an organization can follow is conducting a needs assessment by planning in all areas consulting all departments and creating legal templates for purchase and sale of goods and services. This kind of planning assists in eliminating long wait times for the preparation of legal documents. For example, in terms of legal and corporate affairs, keep prominent positions open for employees who are well aware of the local laws and regulations on mining which assists in cost optimization through in-house counseling and optimizing costs of going in for external legal counseling.

3.3.3 Other Organizational Efficiencies in Mining

Organizational efficiencies in mining also include employee efficiency, process efficiency, performance efficiency, and efficiency in time management. Each of these efficiencies is described in detail below in the mining context through the case study outcomes.

Employee/Manpower Efficiency in Mining

A factor that hinders employee efficiency is that mine operations are usually in remote locations, which is one of the main reasons for skill shortages among employees. Apart from that, in these mine sites, it is also difficult most of the time to find people with good basic skills as mechanics or electricians. The solution to this is to change the education system to include those skills in a technical competency-building curriculum. Management and supervision by managers require them to have sufficient training and experience and it is different from what operators acquire so the training should be accordingly planned for different strata of the organization.

Employee efficiency is related to available resources locally or ex-pat employees hired from outside the country. The cost of hiring local employees is less, they can be trained to acquire certain skills but sometimes if they are not able to acquire certain skills due to certain conditional factors like work culture locally, then this can affect productivity negatively. It is advisable to hire expertise from outside the country even if it costs more which is a trade-off to improve productivity.

Process Efficiency in Mining

Like in every other organization, process optimization is an important factor in mining. All areas in mining whether it is operations, processing, finance, or administration, mining companies need to identify the various constraints and bottlenecks that make the process longer or have unnecessary processes added to it and then use scientific methods to eliminate the weaknesses of such processes.

Performance Efficiency in Mining

In mining, employee performance is generally tied up to remuneration in terms of incentives, and bonuses and management executives are rewarded in terms of stock options when they achieve their performance targets. The targets are based on key performance indicators or KPIs. Many mining organizations put KPI's to every department making it easier to measure the performance of all resources by relating to employees and the KPIs put to other resources. KPI metrics can be instantly visualized, tracked, and analyzed through information systems.

Time Efficiency in Mining

In mining, eliminating wait times is an important factor whether it is about an equipment, which is available but is waiting to be utilized, or personnel waiting to be assigned a task, or departments waiting for spare parts or components to be procured. Each of these activities required to be monitored to eliminate the wait times which otherwise will incur costs. Time factor must be considered very closely in terms of understanding where time has been wasted by using techniques like Pareto analysis. It can be something as simple as the quality of mine roads, which relates to hauling or lading times. Even a wastage of 10–15 s in loading can cost a lot when each loading cycle per day is considered.

As an example, in blasting, if seven or eight shots need to be fired in a timeframe of 30 min and identification of saving 10 min and doing it within 20 min would save a lot of capital, since during the blasting timeframe operations is also stopped and this means saving the 10 min would allow operations for an extra 10 min, which is a direct time leading to cost saving. Cycle times play an important role in operations, in case of loading time, two factors are important loading time for operator and spot time for trucks. A skilled operator can manage to reduce the cycle time considerably which affects productivity positively.

3.4 Equipment in Mining

Equipment in manufacturing companies is complex and requires personnel who are skilled to do their maintenance, which leads to an increase in maintenance costs, which is an important direct cost (Sabaei et al., 2015). Maintenance is an activity that maximizes the capacity of the equipment to produce by affecting the availability and reliability of the equipment which has a direct impact on profitability (Sabaei et al., 2015). The mining environment is challenging for the equipment that is used in large-scale operations and requires innovative and modern approaches to designing the mining equipment (Yudelman, 2006b).

3.5 Production Equipment Used for Large-Scale Open-Pit Mining Operations

Mining fleet equipment selection is an important driver to ensure the safety and profitability of LSOPM operations. A thorough study of mining fleet planning and equipment selection becomes critical to ensure the lowest cost per tonne of material moved over the life of the fleet and the life of the mine to achieve the planned production rates effectively. It is important to note that in an LSOPM operation almost every piece of mining equipment is very much capital intensive. According to the case study outcomes, the primary factors that decide the type, size, and capacity of the mining equipment are:

- Type and size of the deposit
- Mine plan
- Annual and Life of Mine Production capacity
- Capital
- Safety and compliance with regulations

There are three main categories of equipment fleet in mining operations—main production equipment, ancillary equipment, support equipment. Mining and ancillary fleet selection are done taking into consideration the mining schedule, drilling requirement based on the rock type, rock hardness, volume of drilling required to achieve the production plan, blasting requirements, type of explosive used, size and volume of blasts, average haul distances between the loading and dumping points, average truck speeds and cycle time, shift schedules, assumptions on the equipment availability and utilization.

Main Production Equipment

- Drilling and Blasting Equipment
  - **Blasthole Drills** are equipment used for drilling a hole of various diameters and depth in strata or rock mass to facilitate the charging of explosives to break the rock and hard minerals to smaller and optimum fragments to get the waste rock and ore resources to be mined.
  - **Blast-hole Emulsion Explosive Trucks** are bulk explosive loading trucks used to make or charge bulk emulsion explosives ready for pumping into the drilled blast holes for basting the rocks to smaller fragments suitable for loading and transportation.
- Loading and Hauling Equipment
  - **Draglines** are bulk mining primary production equipment using a bucket, pulled by a cable for loading operation. Draglines are most suitable for dismantling and loading of softer materials like loose soil, strata, and ores of very less hardness in a mining operation.

- **Electric Rope Shovels** are electrically powered loading equipment with a rope attachment used to remove large amounts of blasted and fragmented overburden or ore efficiently in open-pit mines.
- **Hydraulic Mining Shovels** are hydraulically operated loading equipment used for digging and loading of fragmented rock or ore to a truck or hauling equipment in open-pit mines.
- **Off-Highway Trucks** are huge mining trucks with a rigid frame or articulated and specially designed to work in harsh mining and construction environments to transport material, overburden, ore, and rocks from the loading point to the dumping point in an open-pit mine.
- **Articulated Trucks** are generally medium-sized off-highway trucks with a permanent or semipermanent pivoting mechanism used to haul a wide spectrum of material in mining, construction, and quarrying operations. They are capable of maneuvering through smaller turning radiuses and steeper gradients.
- **Wheel Loaders** are flexible hydraulic equipment with a bucket attached for loading, lifting, pushing of overburden, rock, ore, sand, and other materials in a mine, quarry, construction, or on-site operations.

3.6 Ancillary Equipment

- **Track mounter dozers** are large hydraulically operated equipment mounted on tracks with a blade and/or ripper attachment used to push large quantities of material, prepare drilling faces, maintain pit and dump floors, prepare safety berms, and rip loose material for loading operations in a mining operation.
- **Wheel Mounted dozers** are hydraulically operated equipment with an articulated actuation mounted on wheels with a blade attachment to push materials, clean roads, maintain pit and bench floors. It is fast and highly maneuverable.
- **Motor Graders** are hydraulically operated equipment with a long blade attachment that is used for leveling the road surfaces and maintenance of the mine roads. Motor graders are also sometimes fitted with a light-duty ripper attachment to support in the ripping and leveling of roads.
- **Wheel Tractor Scraper** is a hydraulically operated equipment with a scraper blade and a hopper that is used to load, transport, and fill the area where the blade gets raised. The rear part of the scraper has a vertically moveable hopper with a sharp horizontal front edge, which can be raised or lowered. The front edge cuts into the soil, and when the hopper is full it is raised, closed, and the scraper can transport its load to the fill area where it is dumped. Scrapers are used to level uneven soft and medium-hard floors.
- **Dust Suppression Water Trucks** are used for dust control in open-pit mining by spraying water to suppress dust for smooth mining operations to ensure the safe movement of people and heavy mining equipment in the mining area.

- **Tire Handler** is a hydraulically operated equipment with a tire handling mechanism to provide a safe and efficient way of changing, mounting, and dismounting of large size heavy duty tires in mining operation.

Support Equipment

The support equipment is required to support the main mining production fleet and ancillary equipment fleet to sustain safe and productive operations. The following is a list of important supporting equipment.

- Crane
- Fire Truck.
- Maintenance truck
- Service and lube truck
- Fuel truck
- Forklift
- Welding truck
- Explosive van
- Pick-up truck
- Crew bus
- Lighting plants
- Dewatering pumps
- Diesel Generators
- Aggregate crusher for blast hole stemming and road building.

3.6.1 Processing Equipment Used for LSOPM Operations

Major equipment used for mineral processing operations are the following:

Crushing Equipment

Crushing of the ROM (Run of Mine) ore is generally done in three stages depending on the type of ore and plant requirements i. primary crushing, ii. secondary crushing, and iii. tertiary crushing. In these three stages, the ROM ore size is reduced from 1.2–1.4 m to less than 0.5 cm. The major crushing equipment includes the gyratory crushers, the jaw crushers, the standard cone crushers, the short head cone crushers, and the roll crushers.

Grinding Equipment

Grinding of the crushed ore is done to achieve the correct degree of size reduction to liberate the interlocked mineral particles from the ore and provide a larger surface area for the mineral processing reagents to react with the mineral particles for liberation. Grinding is done in two stages that are coarse grinding and fine grinding. Coarse grinding reduces the crushed ore particle size from 0.5 cm to about 300 microns and fine grinding reduces it to about 100 microns. The major equipment used for coarse grinding is SAG (semiautogenous) mill or rod mill and fine grinding is done using ball mills.

Screening Equipment

Screening is the process of mechanically separating the ore into various grades by their particle size to obtain the required size for liberation. The major types of screening equipment are grizzlies, trommels, and moving screens like reciprocating screen, vibrating screen, oscillating screens, and gyratory screens.

Classification Equipment

Classification is the separation of mineral particles in a medium into two or more products based on their settling velocities in water, air, or fluid in which they are suspended. The major types of classifiers are hydraulic, mechanical, and cyclones classifiers. Hydraulic classifiers are generally large pools, ponds, or tanks with a conical bottom with free settling zones. Mechanical classifiers are spiral, or rake classifiers and cyclones are hydro cyclones using centrifugal force to enable the settling of particles.

Ore Concentration Equipment

Ore concentration is the separation of the valuable mineral from the gangue material. The major ore concentration processes are hydraulic washing, gravity separation, magnetic separation, electrostatic separation, optical separation, floatation, and leaching. The major equipment used is hydraulic washing machines, jigs, shaking tables, Knelson concentrators, electro-optic detectors, magnetic rolls, low intensity, and high-intensity magnetic separators, leach tanks, and floatation cells.

Filtration Equipment

Filtration is the process of separating the fluid with suspended solids into a solid filter cake and liquid filtrate by passing it through a permeable filter membrane. The major filtering equipment is gravity filters like screens, centrifugal filters like centrifuges, vacuum filters like drum filters and disk filters, and highly efficient filter presses like pressure filters which can drop water levels to about 10% moisture content in the concentrate.

Thickening Equipment

Thickening is the process of settling of suspended solids in a tank under the influence of gravity to form a pulp. The major thickening equipment is mechanical thickeners using rakes that move the material settled at the bottom to the discharge point which can then be pumped using high capacity pumps. Flocculants are chemicals added to assist the thickening process.

Drying Equipment

Drying is the process of removal or large amounts of water from the concentrated ore. Dryers generally use a flow of hot combustion gases to remove moisture from a stream of pulp or concentrates. The major drying equipment is rotary drum dryers, flash dryers, spray dryers, conveyor dryers, and fluidized bed dryers.

3.6.2 *Availability and Utilization of Mining Equipment*

The increased usage and demand for metals cause demand for mining machinery to rise, which in turn causes top mining equipment manufacturers to account for about one-fourth of total production globally (Deneen & Gross, 2009). In terms of machinery production, the major production of mining equipment is done by countries like China, USA, Germany, the UK, and Australia, followed by Brazil and South Africa, and countries like India are also making their way to be major producers of mining machinery (Deneen & Gross, 2009).

The demand and supply of mining equipment on a regional basis are high in Africa, Asia–Pacific, and Latin America rely on mining, but these nations cannot manufacture sophisticated equipment (Deneen & Gross, 2009). Industrialized countries in Western Europe, Japan, South Korea, and Taiwan have small mining sectors but manufacture very high-end mining equipment to be exported and used by companies around the globe. Similarly, developed countries like Australia, Canada, South Africa, and the USA are significantly involved in mining activities and they manufacture the advanced mining equipment, while, Brazil, China, and Russia though have a large presence in natural resource sectors do not produce quite so sophisticated mining equipment (Deneen & Gross, 2009).

Mining companies that use the equipment find it easier if the sales/service network operated by the manufacturer or its dealer network see to it that there are no maintenance problems and replacement parts. Mining equipment manufacturer Caterpillar operates in more than 50 countries making after sales service easier for the companies using the Caterpillar equipment and placed near to the 50 locations (Deneen & Gross, 2009). Equipment selection is a critical process of purchasing suitable equipment in surface mining and the equipment purchased must be compatible with the working environment and other operating equipment. Equipment selection for surface mines is done using different types of methods that include methods like optimization, simulation, and artificial intelligence (Burt et al., 2011).

Equipment planning and selection relate to knowing ahead about even the bench height deals with the purchase of equipment like the back-haul excavator, haul-trucks, dozers, and support equipment plus all the drilling machines. For example, if a decision is made to go in for a 15-m bench height, and if a front shovel that reaches 14 m is selected to be purchased since the blast will end up in 15 + about 3 m usually it would then requires two times of loading and hauling of the materials because the equipment can reach only up to 14 m or so. This makes the use the equipment two times for 9 m each or any other combination, which is utilizing the equipment under its capacity thus affecting the equipment efficiency.

Many times, the decision to purchase equipment lies with cost trade-offs with operational time, which requires a comparison between inexpensive equipment and well-branded equipment. The inexpensive ones may require heavy maintenance within a short duration of time while the branded ones come with annual maintenance contracts and bring down maintenance costs for a longer period. But if the CAPEX or capital expenditure of a mining project is less and the LOM is less then going in for a piece of inexpensive equipment is advisable in this case. For example, if the initial operation is for 2 years, the company for going in for less inexpensive equipment, however, when new plans suggest a feasible life of mine of more than 20 years, the organization should plan and procure equipment which may be expensive but in the long term costs less. The price of the mining machinery is determined by the price of the various commodities that use metal and minerals, thereby increasing or decreasing the costs for mining companies. The factors that increase the price of mining equipment, and hence the costs in mining, include trends in inflation locally, import tariffs, customs duties, availability of supplies, and local demand. Financing for equipment is a major decision area for mining companies since mining equipment is quite expensive.

Mining equipment is generally operated under severe conditions stressing a machine's electrical, hydraulic, and mechanical components to the limit of their design capabilities. Therefore, it is a challenge for equipment manufacturers, operations managers, and maintenance managers to maintain high levels of productivity for the equipment. The productivity in terms of equipment like drills, trucks, shovels, and load-haul-dump depends on factors like the machine's inherent design, operator skill, operating environment, and servicing requirements (Roman, 1999). The three concepts of availability, utilization, and capacity are illustrated in Fig. 3.4 (Lewis & Steinberg, 2001).

After the procurement of equipment, its availability characteristics are already determined, and it is up to the mining personnel to operate and maintain the equipment in a manner that will allow them to achieve maximum performance (Roman, 1999). The performance of equipment is an important factor in maximizing the capacity of an organization and when a piece of equipment is unable to perform according to its standard performance, it is in a state of functional failure (Roman, 1999).

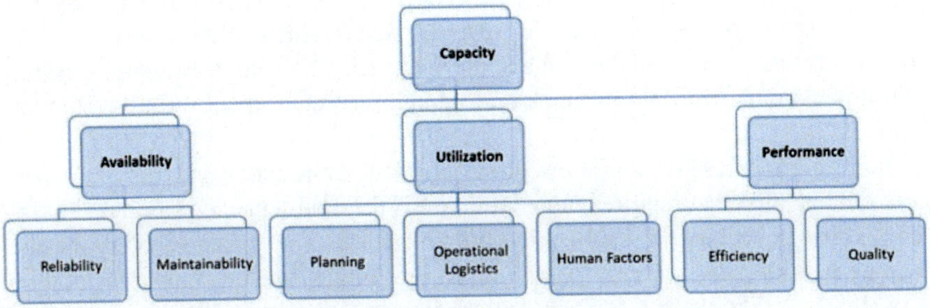

Fig. 3.4 Factors influencing capacity

The terms availability, utilization, and capacity cannot be taken as the same in the case of mining equipment. The availability of equipment is not the same as the utilization of the equipment because a piece of equipment might be available but may not be utilized due to various reasons like operator unavailability, etc. The capacity of an organization can be defined as, "the maximum production capacity of a system under specified operating conditions" (Roman, 1999).

Mining companies decide on having equipment with larger capacities more than equipment with lower capacity. As an example, if a company has 30 haul trucks of 240-tonne capacity (7200 tonnes in total), it is the same as having 20 haul trucks of 360-tonne capacity (Roman, 1999). Having a small fleet of equipment with higher capacity is advantageous since more material can be transported, production costs are minimized and dependency on operators is less.

To understand availability and utilization let us take an example, for a month taken as 24 h a day and 30 days, the working equipment is available for use to say 720 h. If the equipment requires maintenance work and daily check of about 120 h, the availability of the equipment reduces to 600 h, without consideration of breakdown hours due to accidental damage or other reasons. However, there are other factors like shift change hours, lunch break hours, no material available hours, etc., adding up to 40 h which when deducted will reduce the hours to about 560 h, which is the utilization capacity of the equipment out of the available hours. The actual utilization can be compared with the target utilization of the equipment to derive the efficiency of the equipment per month.

A grave disadvantage is that one 360-ton equipment failure is not the same as one 240-ton equipment failure and this leads to productivity losses and an increase in maintenance costs (Roman, 1999). To increase revenue, mining companies are often under pressure to increase production that requires more equipment and machinery time (McRoberts, 2016). Due to tough economic conditions, mining personnel is not able to justify the machine requirements and eventually end up over-using the currently available equipment. This leads to fast equipment failures and reduced the economic life of the machines and their components (McRoberts, 2016).

Operations and maintenance are the two departments in mining organizations that jointly must undertake the responsibility of equipment performance (Lewis & Steinberg, 2001). The operations and maintenance departments need to have a good working relationship, but it is hard to have one when short-range production targets become important in day-to-day management (Roman, 1999). The operations department ends up pushing the limits of the equipment in the high-pressure working environments thus resulting in quick and unpredicted equipment failures (Roman, 1999).

Mining companies give due attention to optimizing production and planning operation, but one area that requires more attention is the maintenance of mining equipment, which contributes a lot to direct mining costs. The repair schedule before equipment failures will largely contribute to cost reduction and improved productivity (Lewis & Steinberg, 2001). A critical concern for mining is the maintenance of the equipment and mining companies need to learn and adapt to new maintenance techniques and best practices in the industry to reduce costs (Peterson et al., 2001c).

An important factor in equipment maintenance is that standard recommendations from OEM are regularly followed in terms of preventive maintenance, which is not enough if the equipment and machinery have increased load or are upgraded (McRoberts, 2016). Unplanned maintenance due to sudden equipment failure can result in loss of production or use of expensive alternatives in comparison to the planned maintenance (Roman, 1999). Maintenance practices can no longer have a conventional approach and mining companies should utilize every available tool and technique to improve them (McRoberts, 2016).

To avoid costly maintenance costs associated with a breakdown, mines adopt preventive maintenance, which means scheduled checkups for equipment to avoid breakdown (Roman, 1999). Maintenance can be divided into:

(a) breakdown maintenance, which is also known as corrective maintenance,
(b) scheduled maintenance, also known as preventive maintenance, and
(c) predictive maintenance.

Predictive maintenance is about techniques available to make predictions about the likelihood of equipment failure over time and maintenance activity can be scheduled based on the predictions to avoid costly failures. Predictive maintenance requires the maintenance personnel to measure the condition of the equipment (Roman, 1999).

One of the important factors is organizational culture that is related to the behavioral aspects in terms of mindsets and attitude to go for preventive measures rather than wait for breakdowns. Many mining companies identified machine reliability as an important factor and started with a reliability-centered maintenance approach. Reliability is about knowing how many times a machine is down in a specific timeframe. For example, in a day a machine is down, and the maintenance team comes, works on it for an hour, and then it is ready for utilization and then in half an hour, it breaks down again. If the number of events occurred for instance in a day is 12 times, the availability is 50% since they lost 12 h out of a total of 24 h a day. Practically KPI monitoring of machine availability alone can sometimes be misleading. A nonreliable machine also causes a production issue. In that case, when you compare productivity, it will be lower than the rated production capacity. Because when the machine is available, then it starts positioning, etc., and then once again when it will be down, it needs to be moved to a safe place, etc. It may be thought of as productivity or production issue, the issue here is more about machine reliability.

For equipment efficiency, the LLF system that means Look, Listen, and Feel can be used in mining organizations. It assists in preventive maintenance practices through visual inspection, sound assessment, and finding variations in the vibration of equipment. It can be used to audit all equipment conditions, monitor all parameters like temperature, pressure, vibration, and visual observations, and helps to reduce the maintenance budget and maintain high operational efficiency.

The concept of the hot seat can be used in the mine operations to improve time management and wastage in time. Instead of parking the heavy equipment during a shift change, keeping the crew of the next shift ready to continue production by eliminating long wait times for equipment. For example, the person using the equipment requires 15 min to park it after a shift and the next personnel reaches 15 min after that, in which case the equipment is available to use but idle for 30 min. If arrangements can be made for reducing the idle time by reducing parking distance and taking over by the next personnel earlier, then the idle time can be eliminated.

The procurement of spare parts and components requires to be aware of all information of vendors who provide them according to the required specification. It would be advisable to have supplier contracts with a few vendors for the provision of certain spare parts that are available only with a few all the time. Not having the right spare part in the inventory when required can cause problems with the availability of the equipment which ends up in a decrease in productivity.

A scientific approach to maintenance management was used from the 1950s and 1960s and preventive maintenance was advised to reduce failures and unplanned downtime (Roman, 1999). Condition monitoring was introduced in the 1970s and the focus was on techniques that predicted the failure of the equipment based on the actual information on the state of the equipment. Equipment manufacturers like Caterpillar use new and advanced technologies that can collect condition monitoring information (Roman, 1999). A detailed study of decision processes and their consequences is necessary to get the optimal costs and benefits of alternative maintenance strategies.

3.7 Organizational Efficiency in LSOPM Operations

Companies aiming to achieve operational excellence make astute choices about the operating structure and execution strategies that will strongly position them for future success. Mineral deposits, which are raw materials for many of the products we use, are often located in remote and inhospitable locations, which pose a complicated logistical challenge for mining operations (CIM, 2008). Developing an economically viable way to get the required manpower and equipment to such locations and the resulting product out of the location is critical for achieving operational profitability (CIM, 2008). A large-scale open-pit mining operation requires skilled management, demand, financing arrangements, and regulatory approval, which motivates the people involved in it to continue performing (Diekmeyer, 2011).

In mining, there is a huge gap in time between investment and returns due to the initial phases of mining in terms of feasibility study and exploration time which varies between projects (Topp et al., 2008). A mining company's success depends on the number and quality of economic deposits that the company can discover and bring into production (McIlroy, 1999). A deposit is called an economic deposit when its evaluation with a specified set of economic and technological conditions leads to minimum acceptable size and profitability criteria. The minimum acceptable size is arbitrary in nature and the minimum profitability criteria are the rate of return on investment is greater than or equal to the weighted average cost of capital, which includes all costs from the start of the development (McIlroy, 1999).

The economic quality of a deposit is depicted in terms of size, profitability criteria, and the difference between the revenues generated from mine production and all the costs required for the production (McIlroy, 1999). The economic potential of a project is evaluated by studies that consider factors pertaining to the size, shape, and location of the deposit, the type of mining method, forecasted ore content of the deposit, forecasted market prices, and the rate at which the ore is to be extracted (Espinoza et al., 2012). Efficient long-term operational mine plans increase the economic viability of a project or allow prospective investors to look for more economical deposits (Espinoza et al., 2012).

Large mining companies use a combination of exploration and acquisition to meet their corporate objectives (McIlroy, 1999). Metal mining and supply to the market can be divided into three phases (McIlroy, 1999):

(a) the exploration phase, the sequential gathering of information stage,
(b) the development phase, the establishment of productive mining and mineral processing result, and
(c) the production phase, processing of ore from the pit until the depletion of resources.

The production plans in an open-pit mining operation may include the flexibility of providing stockpiles of low-grade ore mined in the earlier years which will be processed later as it becomes economical to do so. This means the material is sent to the processing plant or mill either from mine or stockpile (Asad & Topal, 2011). The demand for the metal/mineral is expected to continue increasing while new discoveries of metal deposits will decrease, which will continue the increase in price for metal/minerals (Dold, 2008). The economic boom from 2000 to 2012 leads to the high demand for metals and many mining companies did not pay attention to processes that were losing value by not being economic in the long run (Fazal, 2014). The decrease in commodity prices can lead to less corporate profits, which in the end leads to mine closure, putting shareholders returns in danger and undermining capital budgets; all of which enables mining companies to proactively curb costs and improve productivity (Deloitte, 2015).

There is now a decrease in both labor and capital productivity in terms of volume and cost and top executives in mining companies have realized that regaining lost productivity will be critical for long-term profitability (Fazal, 2014). Mineral resource management can lead to increased productivity and growth which in turn leads to long-term shareholder value. Productivity can be increased by focusing on improved cost structure and increased asset utilization and growth can continue rapidly if the management focuses on expanding revenue opportunities and enhancing customer value (Cinco, 2014).

Selection of equipment does not have a well-defined process in mining because of various factors like knowledge of pit slopes, bench heights, block sizes, and geometrics, ramp layout, excavation sequences, open-pit layout, and production planning (Bazzazi et al., 2009). The most recent equipment selection methods include EQS, which uses fuzzy logic, fuzzy TOPSIS, and MCDM (Bazzazi et al., 2009). The mining process requires materials to be loaded and hauled into large haul trucks and transported to crushers or waste dumps situated away from the pit (Wolpers, 2013). The conventional approach requires the haul trucks to transport material from the pit to crusher or dumpsite continuously, and this process is a major cost contributing factor in mining operations (Wolpers, 2013). The haul truck in open-pit mining operations went through several changes as it evolved; in the 1970s, haul trucks powered electrically in the 200-tonne class were available and in the 1990s 240-tonne class and the mechanical drive trucks became popular (Fiscor, 2016b). The introduction of the 63-inch tire by the end of the 1990s enabled the truck manufacturers to come up with the 360-tonne haul truck. The shovel manufacturers then started developing larger shovels and currently, manufacturers are working on autonomous systems, where driverless trucks that were guided by GPS take ore from the pit to the place where it needs to be dumped (Fiscor, 2016b).

The modern cost reduction technique as developed by the German company, TKF, uses an innovative transport system that consists of a dump station at the foot of the pit. The system is shown in Fig. 3.5 (Thyssenkrupp, 2017). The transport system uses two skips, one for transportation of material to the top, while the other brings the empty bucket that needs to be filled in and sent by the other skip going up. Such a system reduces the cost and time of making the haul trucks go up and down along the pit roads and improves operational efficiency and productivity (Wolpers, 2013).

The strategy that companies can embrace is to continuously invest in innovation by improving mining intensity through recent technologies in terms of data analytics enhanced drilling systems, etc. while reducing people, capital, and energy intensity (Deloitte, 2015). Mine operations have flourished in competitive advantage due to advancements in technology like pumps help miners to dewater the pits, while for rock breakage dynamite replaced black powder (Fiscor, 2016b). Operational improvements in mining companies can be done further up in the value chain by (Deloitte, 2015):

Fig. 3.5 Innovative transport system in open-pit mining

(a) data integration, which is an integration of systems that change data to intelligence needs to be integrated with the ERP platforms,
(b) supply chain optimization, which strengthens vendor relationships across the supply chain,
(c) improved capital allocation, which ties up capital allocation to strategic priorities, and
(d) working capital efficiency, which frees up operational capital for more productive uses.

Many more examples of technological advancement can be seen in the form of compressed air, diesel engines, and electricity, providing the energy to make an enormous efficiency increase. In terms of hauling, locomotive replaced horses, rubber-tired equipment replaced rail haulage, and today truck haulage is replaceable with in-pit crushing and conveying systems (Fiscor, 2016b). Productivity can be improved if mining equipment has better automation and control systems and some equipment manufacturers are already incorporating human-assisted control systems and improvements in man–machine interfaces (Yudelman, 2006b). The mining machinery sector is progressing and advancing technologically by going in for innovations in the form of applications of electronics, using fuel cells and robotics (Deneen & Gross, 2009).

Technological developments in the form of computers, today, provide humongous data to professionals to make swift and optimal decisions (Fiscor, 2016b). Rio Tinto, a leading mining company, went in for a 3D software RTVis™ which retrieves data in real time from automated trucks and drills operating in Rio Tinto mines and identifies the size, location, and quality of ore (CMJ, 2014). The three-dimensional mapping technology decreases waste and operational costs and improves the efficiency of mining operations and focus on extracting high-value ore (CMJ, 2014).

Case Study Outcomes

The focus of every industry is on competitive advantage to do the best in business than competitors, but the mining industry works differently (Cinco, 2014). Mining companies compete against themselves, domestically and internationally, wherein the mining companies do not influence the prices of the mineral produced. To sustain operations, the only control the management has in mining companies is cost control that is achieved by increasing efficiency and investing annually on state-of-the-art technologies in all key operational aspects (Cinco, 2014).

1. Production: In open-pit mining, production is defined as the volume of extraction of rock or mineral from the earth's surface and processing it. The volume of extraction from mining operations and recovery or throughput in the processing mill have both been identified as the definition of production in terms of tonnes extracted or metal produced. A list of all possible outcomes to define production in LSOPM operations is given in Table 3.3 in the appendix section at the end of the chapter.
2. Productivity: Productivity is defined as the efficient use of assets or resources and is related to performance. It is the rate of production and acts as a measure of

the effectiveness of production. Productivity is related to efficiency in the use of material, manpower, time, and equipment. It is considered as a ratio of output to time, manpower resources, equipment, cost of production, and cost of resources. In open-pit mining, the definition of productivity has been identified mainly as a ratio of volumes produced in tonnes or ounces to factors of input costs in terms of manpower, equipment, and time. It is equated to efficiency and has also been related to the economic viability of the project. A list of all possible outcomes to define productivity in LSOPM operations is given in Table 3.4 in the appendix section at the end of the chapter.

3. Cost optimization: Cost optimization was a term described same as cost reduction by two respondents and different from cost reduction or saving by seven respondents, while one of the respondents felt that cost optimization is related to cost reduction, and one of them even related that costs are not to be optimized in an organization. The definition of cost optimization, however, is defined as maximum productivity with minimum resources or the same output at a lower cost. It is related to productivity, performance, and efficiency in spending. Cost optimization is an on-going process and relates to an increase in production with the same or fewer resources in which profit and revenue are important. A compilation of the most common definitions of the terms production, productivity, and cost optimization is given in Table 3.1. The relationship of costs with cost optimization is given in Fig. 3.6.

Cost optimization is related to time efficiency, process optimization, automation, and quality of the output. It requires production efficiency and maximizing net

Table 3.1 Compilation of definitions of terms

Term	Definitions
Production	Volumes produced as per defined targets, key performance indicators, business objectives, and budget of the company. Generates revenue and profit margin and should be economical. Well organized through processes and related to safety, productivity, and cost-efficiency. Related to the extraction and processing of a product
Productivity	The efficient use of assets or resources and related to performance. It is the rate of production and acts as a measure of the effectiveness of production. It can be measured as the ratio of: a. Output to time b. Output to manpower resources c. Output to equipment d. Output to the cost of production e. Output to the cost of resources f. Revenue to costs
Cost optimization	Relates to maximum productivity with minimum resources or the same output at a lower cost. It is related to productivity, performance, and efficiency in spending. Considers cost from the overall business standpoint and relates to an increase in production with the same or fewer resources. It is a continuous on-going process in which profit and revenue are maximized

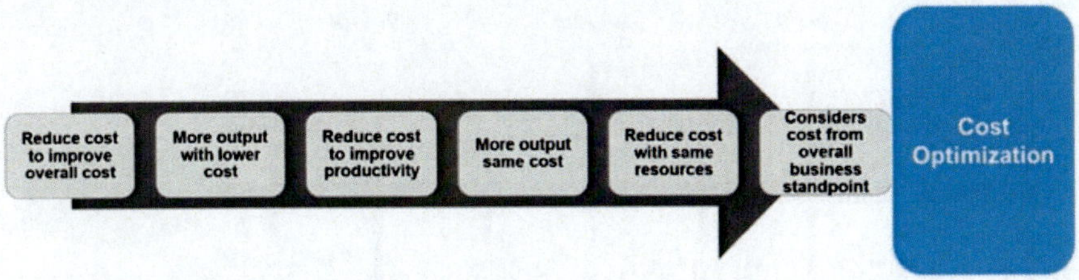

Fig. 3.6 Relationship of cost optimization with cost

present value by defining best practices for the organization. Among other things, cost optimization is also related to productivity, safety, and the environment. The market prices for minerals and metals are independent of variables that cannot be directly controlled by businesses and companies, rather it depends on the local and international market conditions, the only thing that companies can do is to control and optimize costs. A list of all possible outcomes to define cost optimization in LSOPM operations is given in Table 3.5 in the appendix section at the end of the chapter.

3.7.1 Factors for Improving Productivity

The most important factors of productivity identified are efficiencies in manpower resources, equipment, and time. Among other factors, digitization or automation, effective planning and designing throughout the process, and development of employees through training, are some other factors that improve productivity. Manpower efficiency is related to equipment efficiency and is understood by many in terms of a wide variation in the personnel's skill sets and a very good operator will have a very good loading unit time and keeps the trucks in a good cycle.

In terms of manpower resources, employee qualification and attitude play a vital role in productivity in LSOPM operations. As an organization, the LSOPM companies should focus on safety, planning and design, performance measurement, risk assessment, optimizing the entire value chain by ensuring the sustainability of processes to achieve optimum business objectives. Some other factors that affect the productivity of the LSOPM operations are given in Fig. 3.7. A list of all possible outcomes for factors improving productivity in LSOPM operations is given in Table 3.6 in the appendix section at the end of the chapter.

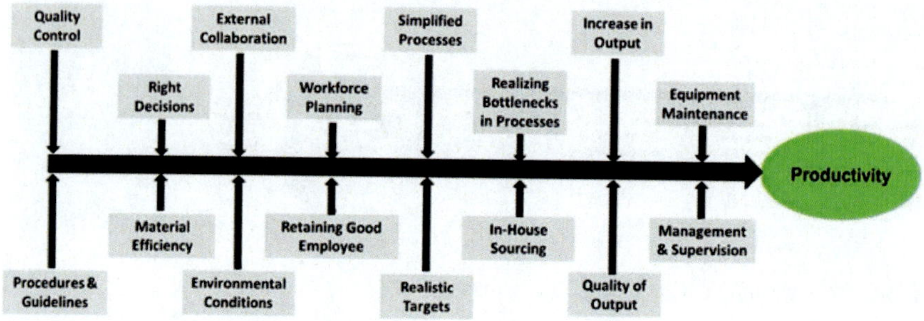

Fig. 3.7 Factors improving productivity

3.7.2 Key Drivers of Cost Optimization

The key drivers identified for cost optimization for LSOPM operations are efficiencies in manpower resources and equipment, improvement in productivity to achieve operational excellence, and process optimization. Process optimization can be achieved in many ways, for example, the management of fleet in terms of how the group is managed, how the different processes are optimized in the same department, how equipment operations are optimized, how daily plans are optimized, and how cycle times are optimized.

A wide list of variables can be determined apart from the above key drivers which include equipment maintenance, safety and environment, use of information technologies, developing and using local resources, measuring performance, and having a good understanding of costs since mineral prices depend on market variation. Other drivers of cost optimization are given in Fig. 3.8.

The key drivers of cost optimization and the factors to improve productivity had many similarities. The relevant similarities were in terms of an increase in production, effective planning, employee training, retaining good employees, workforce planning, and efficiency in time, and quality of the material. However, the differences were

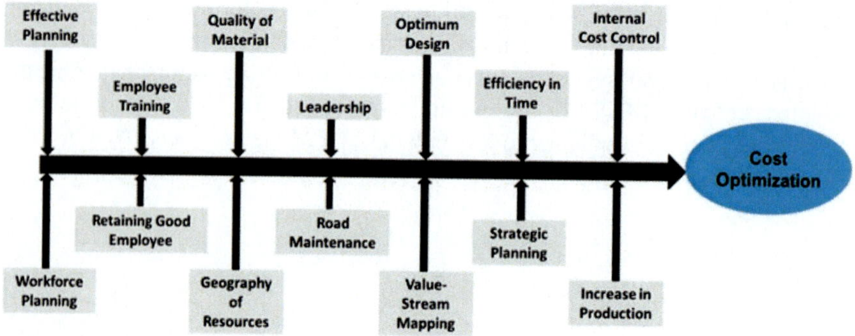

Fig. 3.8 Other key drivers of cost optimization

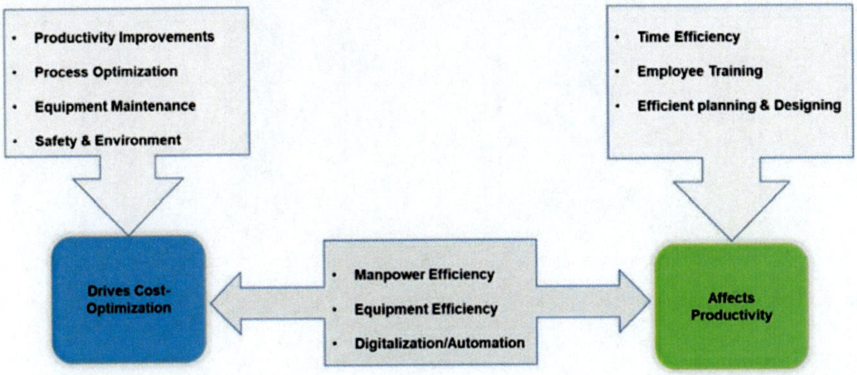

Fig. 3.9 Similarities and differences between key drivers of cost optimization and factors improving productivity

in terms of road maintenance, strategic planning, the geography of resources, value stream mapping, leadership, internal cost control, and optimum design. Figure 3.9 indicates the most relevant similarities and differences between key drivers of cost optimization and factors for improving productivity. A list of all possible outcomes for key drivers of cost optimization in LSOPM operations is given in Table 3.7 in the appendix section at the end of the chapter.

3.7.3 Relationship Between Productivity and Costs

The relationship between productivity with the cost is complex in certain situations when an increase in productivity in one area can result in a decrease in costs in another area. The terminologies that were used repeatedly to understand the interrelationship between productivity and cost optimization were in terms of decrease and increase of costs, resources, output, productivity, time, production, and cost optimization. The relationship between productivity and cost optimization depends on various factors. The principal drivers are the rate of production and unit cost of production.

Although both productivity and cost optimization are related to efficiencies, an increase in productivity generally ensures the optimization of costs across the value chain. Costs can be monitored from time to time by measuring productivity. Manpower efficiency in terms of deployment of resources at the right place, development of skills and training can optimize costs and result in cost optimization. Similarly, efficient safety techniques and environmental care, along with minimum disruption to processes, are some other factors where costs can be optimized, and together changes in productivity can be seen.

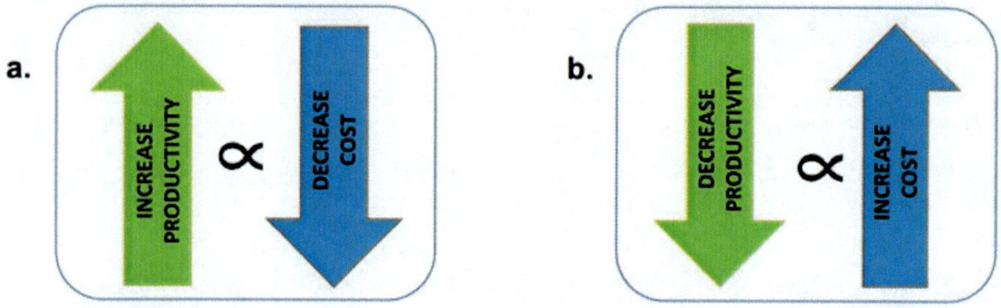

Fig. 3.10 Relationship between cost and productivity

Productivity can be improved by reducing costs appropriately in terms of using fewer resources and achieving more output. The decline of demand in commodities requires improving productivity which in turn requires appropriate cost-cutting measures. Thus, productivity and costs are inversely related and an increase in productivity causes a decrease in costs. The interrelationship between productivity and cost is clearly depicted in Fig. 3.10. The effects of productivity on costs are clear that of inverse proportionality in terms of increase and decrease, which means if productivity is not higher then cost will go up, for example, if there are long cycle times then productivity is not high, it is low so that is going to cost more money. Thus, always when productivity increases cost is low, and when productivity decreases cost is high.

However, there were a few examples in which the general relationship between the variables can be contested. The reduction in cost needs not ensure improvement in productivity. A change in cost needs not affect cost optimization or productivity. The very notion of the cost being optimized in one area is not right, it should never be optimized in one area, so it cannot be directly related to productivity in that area but across the value chain to have an overall business impact. Thus, it can be concluded that although productivity increase and the decrease can be causally related to cost decrease and increase, respectively, the vice versa is subjected to different conditions and circumstances. A list of all possible outcomes for the effects of change in productivity on cost optimization in LSOPM operations is given in Table 3.8 in the appendix section at the end of the chapter.

3.7.4 Role of Equipment Utilization in Productivity

To understand the role of equipment utilization in productivity, it is very important to understand the meaning of equipment utilization, which is also related to equipment availability, use of the equipment in the available time in production, general maintenance practices, and most importantly the selection and procurement of the appropriate equipment. The process starts with the selection of equipment. For the planning and success of the operations, it is important that the selection of the equipment is right in terms of specifications and use, failing to adhere to it would result in inefficiency of operations. The selection of equipment also depends on the suitability of the equipment for that production requirement, operating cost of production, the capital cost of the equipment itself, and the productivity of the equipment.

The terms availability, utilization, and capacity cannot be taken as the same in the case of mining equipment. The efficient maintenance of the mining equipment in LSOPM operation results in an increase in productivity. The efficiency in the maintenance of equipment can be increased with the aid of information technologies and the positive attitude of employees in understanding that the equipment must be operated efficiently. The reliability factor of the equipment is also an important factor for equipment maintenance, which helps in understanding at what rate equipment has breakdowns and if it is frequent, then proactive measures need to be taken.

Equipment availability depends on good maintenance practices like for example, availability of backup equipment in process plants increases utilization and it can sometimes be equal to equipment availability. However, backup is not possible always for the production operations and this means the dependency is a lot on how the maintenance of equipment takes place. Equipment availability is basically, equipment being available for use and if it is low in availability, it will affect production. It also depends on the right equipment being used for the operations. Figure 3.11 shows the main three factors on which equipment utilization depends.

Equipment availability does not always ensure equipment utilization. In some instances, the equipment might be readily available but may not be in the right place to be utilized. Equipment utilization is the efficient usage of the equipment when it is available. The gap between the equipment availability and utilization needs to be reduced, the lesser the gap, the more efficiently the equipment is in use while available. The maximum utilization of equipment maximizes equipment

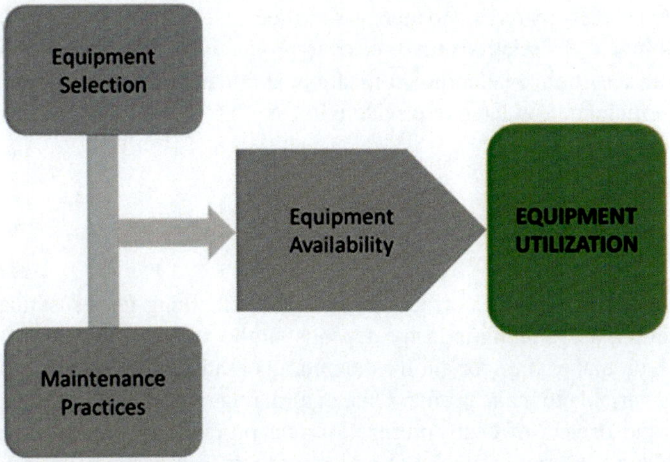

Fig. 3.11 Equipment utilization

Table 3.2 Factors improving productivity versus factors improving equipment utilization

Factors improving productivity	Factors improving equipment utilization
Manpower resources: Employee training Employee attitude Employee skill	Manpower resources: Employee training Employee attitude Employee qualification Employee efficiency
Depends on equipment maintenance	Depends on equipment maintenance
Depends on minimizing time	Reducing equipment idle time increases productivity
Depends on performance measurement	Utilized according to benchmarks and key performance indicators
Automation/digitization	Information technologies help in optimizing fleet utilization

productivity. This can be further understood from Table 3.2, which shows the factors that are similar for improving both equipment utilization and productivity. A list of all possible outcomes for the role of equipment utilization on productivity in LSOPM operations is given in Table 3.9 in the appendix section at the end of the chapter.

3.8 Conclusion

This chapter focuses on organizational efficiencies and its relevance in industries. Organizational efficiency is causally related to operational excellence and in this context, a detailed overview of productivity and costs is important for every organization. Productivity and costs along with equipment efficiency play a crucial role in LSOPM operations. The relationship between productivity and costs shows that productivity is inversely related to increase and decrease in costs, however, this is not proven in terms of the relationship of costs to productivity. Equipment efficiency is related to the selection, availability, utilization, and maintenance of equipment, and equipment efficiency is related to productivity.

3.9 Summary

Organizational efficiency is a critical focus area for the long-term sustainability of an organization. The mining industry depends on the market for prices of the end-product, which makes it important for the mining organizations to consider costs as the primary variable to make profits. This, in turn, relates to productivity, equipment efficiency, and drivers of cost optimization and productivity improvements. Key similarities and differences among organizational efficiencies have been identified in this chapter to lead to operational excellence in the LSOPM scenario.

Appendix: Additional Information on Organizational Efficiencies in LSOPM Operations

See Tables 3.3, 3.4, 3.5, 3.6, 3.7, 3.8 and 3.9.

Table 3.3 List of possible definitions of production

#	Description
1	Tonnage, ounces, or volume produced
2	Associated with achieving targets, KPI's, objectives, or budget
3	Extracting rock or mineral from earth
4	Generates revenue and profit margin
5	Economical
6	Extracting and processing
7	Well organized through processes
8	Related to safety
9	Related to productivity
10	Done efficiently at a lower cost
11	To produce or to make output from input

Table 3.4 List of possible definitions of productivity

#	Description
1	Efficient use of assets or resources
2	Related to performance- KPI, OEE, and achieving budgeted targets
3	Ratio of output to time
4	Ratio of output to manpower resources
5	Ratio of output to equipment
6	Rate of production
7	Process optimization
8	Measure of effectiveness of production
9	Ratio of output to cost of production
10	Equipment efficiency
11	Optimum profit margin
12	Technology dependent
13	Optimization of time

(continued)

Table 3.4 (continued)

#	Description
14	Material optimization
15	Ratio of output to cost of resources
16	Manpower efficiency
17	Ratio of revenue to costs
18	Related to safety
19	Economically viable planning
20	Related to cost

Table 3.5 List of possible definitions of cost optimization

#	Description
1	Different from cost reduction or cost saving
2	Same output with lower cost
3	Related to productivity
4	Maximum productivity or efficiency with minimum resources
5	Efficiency in spending
6	Related to performance—OEE and KPI
7	Considers cost from overall business standpoint
8	Increase in production with same or less resources
9	Metal market price not controlled by businesses
10	Process optimization
11	On-going continuous process
12	Profit and revenue is more important
13	Reducing cost with same resources
14	Same as cost reduction
15	Considers safety and environment
16	More output with same cost
17	Automation is important
18	Quality of the output
19	Reducing cost to improve overall cost
20	Utilizing assets effectively
21	Related to saving time
22	Defining best practices for your organization
23	Maximizing net present value
24	Similar to productivity
25	Fewer brilliant resources than many low productivity resources
26	Improves productivity

(continued)

Table 3.5 (continued)

#	Description
27	Reduce cost and improve productivity
28	More output with lower cost
29	Identifying bottlenecks to realize value
30	Related to cost reduction
31	Production efficiency
32	Costs should not be optimized

Table 3.6 List of possible factors for improving productivity

#	Description
1	Employee efficiency
2	Digitalization or automation
3	Minimizing time
4	Employee training
5	Equipment efficiency
6	Effective planning and designing
7	Safety
8	Employee qualification
9	Performance measurement
10	Employee attitude
11	Procedures and guidelines
12	Assessing risk
13	Optimizing entire value cycle
14	Sustainability of processes
15	Quality control
16	Quality of output
17	Environmental conditions
18	Increasing output
19	Retaining good employees
20	Right decisions
21	Realizing bottlenecks in processes
22	Realistic targets
23	In-house sourcing
24	Material efficiency
25	Workforce planning
26	Management and supervision

(continued)

Table 3.6 (continued)

#	Description
27	Equipment maintenance
28	Simplifying processes
29	External collaboration

Table 3.7 List of possible key drivers of cost optimization

#	Description
1	Efficient manpower resources
2	Equipment efficiency
3	Improving productivity/operational excellence
4	Optimizing processes
5	Equipment maintenance
6	Safety and environment
7	Information technologies/automation/digitalization
8	Understanding costs
9	Market variation and mineral price
10	Developing and using local resources
11	Measuring performance
12	Road maintenance
13	Increase production
14	Leadership
15	Effective planning
16	Employee training
17	Internal cost control
18	Value stream mapping
19	Geography of resources
20	Strategic planning
21	Management decisions
22	Optimizing design
23	Time factor
24	Workforce planning
25	Retaining good employees
26	Company objectives

(continued)

Table 3.7 (continued)

#	Description
27	Reducing waste
28	Quality of material purchased in low cost
29	Capacity of resources
30	Minimum outsourcing
31	Accountability, responsibility, and strict governance
32	Decreasing production costs

Table 3.8 List of possible outcomes for effect of change in productivity on cost optimization

#	Description
1	Reduce operating cost to increase productivity
2	Productivity and cost optimization are related
3	More output, same resources, increases productivity, and reduce cost
4	Decrease in productivity will increase costs
5	Change in productivity is related to change in costs
6	Measure productivity to monitor increase or decrease in costs
7	Increasing productivity will reduce costs and optimize costs
8	Increase in time increases cost and decreases productivity
9	Decrease in production increases production costs
10	Increase in productivity may not decrease costs
11	Using same resources increase revenue and decrease costs
12	Change in cost should not affect productivity
13	Increase in cost needs not to affect cost optimization
14	Increase in productivity does not ensure cost optimization
15	Improving production does not require reducing costs
16	Manpower efficiency increases cost optimization
17	Disruptions to processes impact unit costs
18	Continuously monitor cost and productivity
19	Costs should not be optimized so not related to productivity
20	Efficient environmental care improves mine life
21	Efficient safety methods will ensure cost optimization
22	Very productive with least resources make cost efficient
23	Educate and train employees to optimize costs
24	Workforce stability reduces cost
25	Productivity less in an area and more in another reduces costs

Table 3.9 List of possible outcomes for role of equipment utilization on productivity

#	Description
1	Equipment utilization is efficient usage of equipment when it is available
2	Equipment utilization affects cost optimization
3	Equipment utilization depends on time efficiency
4	Equipment utilization is done according to benchmarks and KPI
5	Information technologies help in optimizing fleet utilization
6	Equipment utilization depends on equipment availability
7	Effective planning required
8	Training employees important for efficient equipment utilization
9	Optimize processes to improve equipment utilization
10	Reducing equipment idle time increase productivity
11	Maximize equipment utilization to maximize productivity
12	Overall efficiency more important than individual efficiency
13	Equipment utilization has no direct relationship to productivity
14	Employee attitude important for equipment utilization
15	Employees skill is important in equipment utilization
16	Inefficient use of equipment impacts costs
17	Equipment utilization depends on equipment maintenance
18	Decrease in equipment utilization decreases profitability
19	Equipment utilization of a fleet is more important than individual EU
20	If not required reduce equipment use that reduces cost
21	Equipment utilization, equipment availability, and productivity are related
22	Equipment utilization and equipment availability are not causally linked

References

Armistead, C. (1991). *Resource Productivity Management in the Services Sector*. Bedford: Cranfield University.

Asad, M. W., & Topal, E. (2011). November). Net present value maximization model for optimum cut-off grade policy of open pit mining operations. *the Journal of the Southern African Institute of Mining and Metallurgy, 111,* 741–750.

Asif, M., & Searcy, C. (2014). Determining the key capabilities required for performance excellence in higher education. *Total Quality Management*, 22–35.

Baack, D. W., & Boggs, D. J. (2008). The difficulties in using a cost leadership strategy in emerging markets. *International Journal of Emerging Markets, 3*(2), 125–139. https://doi.org/10.1108/17468800810862605

Ball, P. D., Roberts, S., Natalicchio, A., & Scorzafave, C. (2011). Modelling production ramp-up of engineering products . *Journal of Engineering Manufacture*, 959–971.

Bazzazi, A. G., Osanloo, M., & Behrooz, K. (2009). Optimal Open Pit Mining Equipment Selection using Fuzzy Mutiple Attribute Decision Making Approach. *Archives of Mining Sciences, 54*(2), 301–320. Retrieved from https://mining.archives.pl/.

Bigelow, M. (2002, October). How to achieve operational excellence. *Quality Progress, 35*(10), 70–75.

Britney, R. R., Kudar, R. P., Johnston, D. A., & Walsh, J. (1986). Forum/Key Productivity Issues: A Comparison of International Productivity Centers. *National Productivity Review*, 71–76.

Burt, C., Caccetta, L., Welgama, P., & Fouche´, L. (2011). Equipment selection with heterogeneous fleets for multiple-period schedules. *Journal of the Operational Research Society* , 1498–1509.

Chaffey, D., & White, G. (2012). *Business Information Management*. Edinburgh: Pearson Education Limited.

Chevron Corporation. (2010). *Operational Excellence Management System: An Overview of the OEMS*. Retrieved from Chevron Corporation Web site: www.chevron.com.

CIM. (2008, June/July). *Remote Control: The logistical challenges of remote mining operations - See more at:* https://www.cim.org/en/Publications-and-Technical-Resources/Publications/CIM-Magazine/2008/June/features/Remote-control.aspx?page=1#sthash.4MywCdBv.dpuf. Retrieved from Canadian Institute of Mining, Metallurgy and Petroleum: https://www.cim.org/en/Publications-and-Technical-Resources/Publications/CIM-Magazine/2008/June/features/Remote-control.aspx?page=1.

Cinco, C. J. (2014). Philippine Open Pit Copper Operations: Enhancing Operational Efficiency through Value Chain Analysis . *Philippine Management Review*, 69–86.

CMJ. (2012, June 4). *Commentary: Cost overruns on capital development: An ever-growing concern*. Retrieved from Canadian Mining Journal Web site: www.canadianminingjournal.com.

CMJ. (2014, September 23). *SOFTWARE: Real time ore tracking for improved productivity*. Retrieved from Canadian Mining Journal Web site: www.canadianminingjournal.com.

Cosgrove, C. V. (1986). How to Report Productivity: Linking Measurements to Bottom- Line Financial Results. *National Productivity Review*, 63–70.

Dagdelen, K. (2001). Open Pit Optimization - Strategies for Improving Economics of Mining Projects Through Mine Planning. *17th International Mining Congress and Exhibition of Turkey-IMCET2001*, (pp. 117–121).

Dassault Systems GEOVIA. (2017). *Home: Dassault Systems GEOVIA*. From Dassault Systems GEOVIA: www.geovia.com.

Deloitte . (2015). *Tracking the trends 2016*. Deloitte Touche Tohmatsu Limited. Retrieved 2015, from https://www2.deloitte.com/content/dam/Deloitte/global/Documents/Energy-and-Resources/gx-er-tracking-the-trends-2015.pdf.

Deneen, A. M., & Gross, A. C. (2009). World Mining Machinery. *Business Economics*, 169–176.

Diekmeyer, P. (2011, February). *Fast-Tracking to Production*. Retrieved from Canadian Institute of Mining, Metallurgy and Petroleum: https://www.cim.org/en/Publications-and-Technical-Resources/Publications/CIM-Magazine/2011/february/project-profile/Fast-tracking-to-production.aspx.

Dold, B. (2008, September 27). Sustainability in metal mining: from exploration, over processing to mine waste management. *Rev Environ Sci Biotechnol* , 275–285. doi:https://doi.org/10.1007/s11157-008-9142-y.

Donnelly, H. J., Gibson, L. J., & Ivancevich, M. J. (1978). *Fundamentals of Management: Functions, behavior, models*. Dallas: Business Publications Inc.

Driussi, C., & Jansz, J. (2005). Technological Options for Waste Minimisation in the Mining Industry. *Journal of Cleaner Production*, 682–688.

Duggan, K. (2013, October 28). *Achieving Pharmaceutical Operational Excellence*. Retrieved from Advantage Business Media Web site: www.pharmpro.com.

E&MJ. (2012a). Australian Technology: Technology Drives Improved Productivity. *Engineering & Mining Journal*, 102–115.

Espinoza, D., Goycoolea, M., Moreno, E., & Newman, A. (2012, December 6). MineLib: a library of open pit mining problems. *Annals of Operations Research*, 93–114.

Fazal, Z. (2014, October 1). *Productivity challenge: Improving or transforming?* Retrieved from Canadian Mining Journal Web site: www.canadianminingjournal.com.

Fiscor, S. (2016b, September). Tracking 150 Years of Mining Engineering: Problem solvers overcome adversity with advances in equipment and technology. *Engineering & Mining Journal*, 110–156.

Goodfellow, R. C., & Dimitrakopoulos, R. (2016). Global Optimization of open pit mining complexes with uncertainty. *Applied Soft Computing*, 292–304.

Griffith, R., Haskel, J., & Neely, A. (2006). Why is Productivity so Dispersed? *Oxford Review of Economic Policy, 22*(4), 513–525. https://doi.org/10.1093/oxrep/grj030

Hicks, G. H., & Gulliett, R. C. (1981). *Management*. New York: McGraw Hill Inc.

Hiyate, A. (2017). *February)* (pp. 20–25). Pushing Boundaries: A snapshot of mining R&D activity in Ontario. Canadian Mining Journal.

Hustrulid, W., Kuchta, M., & Martin, R. (2013). *Open Pit Mine Planning & Design*. London: CRC Press.

Ion, N., Radu, C., & Georgiana, C. (2015, October 7). Operational excellence—A key to world class business performance. *9th Series of Seminars for Business Professionals*, (pp. 133–140). Paramaribo.

Itanyi, O., Ewurum, F. J., & Ukpere, I. W. (2012, November 7). Evaluation of decision making criteria with special reference to quantitative and qualitative paradigms. *African Journal of Business Management, 6*(44), 11110–11117.

Johnson, T. B. (1968). *Optimum open pit mine production scheduling*. Berkeley: Operations Research Center, University of California.

Kanghwa, C. (2010). From operational efficiency to financial efficiency. *The Asian Journal on Quality, 11*(2), 137–145. https://doi.org/10.1108/15982681011075943

Krol, M., & Brouwer, W. (2014, February 7). How to estimate productivity costs in economic evaluations. *PharmacoEconomics*, 335–344. doi:https://doi.org/10.1007/s40273-014-0132-3.

Lala, A., Moyo, M., Rehbach, S., & Sellschop, R. (2016, August). *Productivity at the Mine Phase: Pointing the Way Forward*. Retrieved from McKinsey & Company Web site: www.mckinsey.com.

Lewis, M. W., & Steinberg, L. (2001). Maintence of mobile mine equipment in the information age. *Journal of Quality in Maintenance Engineering*, 264–274.

Mann, R., Adebanjo, D., & Tickle, M. (2011). Deployment of business excellence in Asia: An exploratory study. *International Journal of Quality & Reliability Management, 25*(6), 604–627. https://doi.org/10.1108/02656711111141184

Mast, J. d., Kemper, B. P., Wiltjer, A., & Does, R. J. (2013). Quality quandaries: Deployng operational excellence at a financial service provider. *Quality Engineering*, 298–306.

Mayhew, K., & Neely, A. (2006). Improving productivity—Opening the blackbox. *Oxford Review of Economic Policy, 22*(4), 445–456. https://doi.org/10.1093/oxrep/grj026

McIlroy, A. R. (1999). *The return from exploration success: Relating economic quality to geological quality*. Ann Arbor: UMI Publishing.

McRoberts, P. (2016, September). World mining equipment: Operating strategies-counting on the connected mine. *Engineering & Mining Journal*, 158–164.

Miguel, S. A. (1996). *Cost components and methods employed in a feasibility study for a typical open pit and underground mining operation*. Ann Arbor: UMI company.

Mikesell, R. F., & Whitney, J. W. (2016). An Overview of the world mining industry. In R. F. Mikesell, & J. W. Whitney, *The world mining industry* (Vol. 11, pp. 1–21). New York: Routledge: Taylor & Francis Group.

Mitchell, P., & Steen, J. (2017). *Productivity in mining: A case for broad transformation*. Retrieved from Earnst & Young Global Limited: http://www.ey.com.

Peterson, D. J., LaTourrette, T., & Bartis, J. T. (2001c, May). Rand report: Third installment: Critical technologies for process optimization. *Engineering & Mining Journal*, 46–52.

Roman, P. A. (1999). *Joint enhancement of mine operations and maintenance: A structured analysis approach*. Queen's University (Canada). Ann Arbor: ProQuest, UMI Dissertations Publishing. Retrieved from https://search.proquest.com/business/docview/304542219?accountid=150425.

Rusev, S. J., & Salonitis, K. (2016). Operational excellence assessment framework for manufacturing companies. *5th CIRP Global Web Conference Research and Innovation for Future Production* (pp. 272–277). Elsevier B. V.

Sabaei, D., Erkoyuncu, J. A., & Roy, R. (2015). A review of multi-criteria decision making methods for enhanced maintenance delivery. *Procedia CIRP*, 30–35.

Schiff, B. J. (2014). Building a sustainable cost leadership culture. *Strategic Finance*, 47–51.

Singh, V. P. (2012, June). Managerial decision making. *International Journal of Innovative Research in Commerce & Management, 5*, 1-4.

Steffen, O. K. (1997). Planning of open pit mines on a risk basis. *The Journal of the south African institute of mining and metallurgy*, 47–56.

Thyssenkrupp. (2017, November 28). *Products and services/mining-systems: Thyssenkrupp-industrial-solutions AG.* Retrieved from Thyssenkrupp-industrial-solutions AG Website: https://www.thyssenkrupp-industrial-solutions.com/.

Topp, V., Soames, L., Parham, D., & Bloch, H. (2008). *Productivity in the mining industry: Measurement and interpretation.* Productivity Commission.

Valery, W., & Jankovic, A. (2011). New methodology to improve productivity of mining operations. ResearchGate.

Vele, C. L. (2014). An empirical study on how the efficient use of resources influences performance. *Managerial Challenges of the Contemporary Society, 7*(1), 88–91.

Vrellas, C. G., & Tsiotras, G. D. (2014, January). Operational excellence in the Greek brewing industry. *Wiley Periodicals. Inc.*, pp. 31–38.

Wolpers, F. (2013, May). Cost-efficient transport for open-pit mines. *Engineering and Mining Journal*, 48–52.

Yudelman, D. (2006). Mining technology: Scratching the surface. *Canadian Mining Journal*, 5.

Chapter 4
Efficient Decision-Making in LSOPM Operations

Abstract Decision-making plays an especially important role in an organization. This chapter deals with a detailed understanding of general decision-making in terms of types of decisions, individual and group decision-making, decision-making processes and models, strategic decision-making, effective managerial decision-making, and tools and techniques used for group decision-making. Apart from that, the chapter shows that in LSOPM operations, the results of primary research in terms of the advantages of better decision-making between the interrelationship of swift decision-making and productivity. In conclusion, a model showing the 11 factors that are relevant to swift decision-making and productivity is presented in the end.

Keywords Effective managerial decision-making · Decision-making processes · Decision-making models · Faster decision-making · Decision-making and productivity

4.1 Introduction

In today's complex, rapidly changing world, decisions are required to be taken in an organization all the time and it is a choice made from available alternative courses of action (Natale et al., 1995). The term business enterprise is often associated with large companies where decisions are made by managers (Bridge & Dodds, 2018). The management has the following function in an organization: (a) decision-making and policy formulation, (b) planning and controlling, (c) organizing and staffing, and (d) communicating and directing (Bridge & Dodds, 2018). Many of the management activities in organizations are unsystematic and managers regularly are faced with making decisions on problems that have rules that are not well defined (Anthony & Govindarajan, 2007). To make management actions systematic, it is required to have proper management control in an organization. Management control is the implementation of the organization's strategies through the influence of managers on other members (Anthony & Govindarajan, 2007).

Fundamentally, management processes have three elements: decisions, managers, and results (Anthony et al., 2014). A decision is a judgment to be made by choosing one option from many alternatives and is not about choosing between right and

H. Qudrat-Ullah et al., *Operational Sustainability in the Mining Industry*, Asset Analytics, https://doi.org/10.1007/978-981-15-9027-6_4

wrong (Anthony et al., 2014). However, it can be elaborated that decisions are sometimes complex, involving multiple, conflicting objectives, where a better outcome for one objective may result in a worse outcome for another objective (Roberts, 1999). Managers today are faced with challenges and need to prove themselves with their capabilities, characteristics, knowledge, and intuitions, which they provide as an input to the organization to gain management control (Anthony et al., 2014). Managers make decisions and are held accountable for the outcome of the decision. Three important factors influencing managerial decision-making in an organization are (a) management skills and approaches towards decision-making, (b) decision supporting the role of information system along with performance measurement and management practices, and (c) attitudes of top management toward stakeholders (Paprika et al., 2008).

Managers must take into consideration how a decision affects the organization as a whole and not just be concerned with a department or area of management (Bridge & Dodds, 2018). One of the characteristics of circumstances of decision-making is that the decisions are riskier as it goes up to a higher level of management. The risk depends on the level of management and the type of decision taken. For example, at the lowest level of management, the most routine type of decisions will have the least degree of risk (Sopta & Slavica, 2017). Decision-making in organizations involves managers who often find themselves burdened with several alternatives to a problem that requires them to apply a realistic approach in choosing and implementing optimal decisions (Itanyi et al., 2012). The efficiency of a manager in decision-making relies largely on the availability of the required information and the experience in handling similar situations (Sopta & Slavica, 2017).

Systems, business planning, and decision-making of managers should all be integrated to achieve optimum efficiency in large organizations (Fiscor, 2007). In the mining area, various decisions like production targets, crew allocation, and equipment allocation are some of the critical issues that require the constant attention of managers (Van Niekerk, 2013). The performance level of key indicators in all areas of mining should be closely monitored and swift decisions need to be taken at the right time to save cost and time (Fiscor, 2007). Due to the increase in global competition, there is uncertainty in terms of being exposed to more competitors and more markets because of which the quality of decision and timeliness of decision-making is essential for every organization (Meagher & Wait, 2010; Singh, 2012). This chapter is divided into three sections:

- Decision-making levels
- Efficient decision-making processes
- Decision-making in LSOPM operations

4.2 Decision-Making Levels

Until the 1980s, managers were expected to monitor, supervise, and control, which did not require them to know the ground reality and they depended on reports to make major managerial decisions (Slater, 2003). Jack Welch, from General Electric, believed in the philosophy that managers need to manage less, and the less they manage the better it is for the company (Slater, 2003). However, it can be emphasized that there should be a commitment toward action and results before making decisions and everyone who has a major stake in terms of accountability and responsibility, due to the decision, must participate in the decision-making process (Drucker, 1974).

4.2.1 Types of Decisions

Accountability and responsibility can be categorized into three levels within an organization—the strategic level, the tactical level, and the technical level, which can be also used for categorizing decisions (Pownall, 2012). The strategic level deals with the top management responsible for long-term decisions, the tactical level deals with the middle-level management who work for the decisions made by top management as goals and objectives, and the technical level decisions are those made by first-line supervision (Hicks & Gulliett, 1981; Pownall, 2012). In an organization, decisions also occur at different levels and can be of different types. Decisions in an organization can be categorized into three types: strategic, administrative, and operating decisions (Bridge & Dodds, 2018).

Strategic decisions are bound by time and knowledge where time is related to the survival of the firm while knowledge is related to the information flow, the uncertain world outside the firm, and nonrepetitive decisions on which very little information is available (Bridge & Dodds, 2018). Strategic decisions are the concern of the top management and have a strong effect on the organization's future (Bridge & Dodds, 2018; Haider & Mariotti, 2016). Strategic decisions involve the most appropriate use of resources for a given competitive goal (Bridge & Dodds, 2018). Operating decisions are related to day-to-day current operations and spread downward in terms of the level of management and may sometimes involve routine decisions for which standardized procedures can be made available (Pownall, 2012). Operational decisions are short-term and responsive decisions (Bridge & Dodds, 2018). Administrative decisions are related to organizational structuring for maximum performance. Another kind of decision is the tactical decisions in an organization that relates to the actions taken to achieve the goal as derived by the strategic decision (Pownall, 2012). The types of decisions can be related directly to the levels in the organization, namely top level, middle level, and lower level (Donnelly et al., 1978). The classification of decisions in various studies differs in name and description, some of the more common classifications are given in Table 4.1 (Natale et al., 1995; Roberts, 1999; HIA Ltd., 2015; Donnelly et al., 1978; Hitt et al., 2014; Pownall, 2012; Harrison & Pelletier, 2000).

Table 4.1 Types and classifications of decisions according to various studies

#	Types and classifications of decisions	Description
1	Sequential	A present decision could be taken if the previous decision is known
2	Common	Decisions of the manager that include scheduling, hiring, firing, etc. Common decisions are repetitive and routine-like, requiring less managerial time and most of the time can be made using decision support tools and computer systems
	Uncommon	Decisions of the manager that include changing strategies, opening a new factory, etc
3	Simple	Decision-making on our routine day-to-day work
	Complex	Decisions dealing with risk and uncertainty and incorporation of people issues
	Complicated	Decisions with a calculated number of choices based on available information
4	Programmed	Decisions are a standard response to a routine problem
	Nonprogrammed	Decisions that are responses to problems that are complex
5	Structured	Decisions that are clear and unambiguous
	Unstructured	Decisions that are unclear and ambiguous
6	Category I	Decisions are decisions that can be routine and recur with a certain outcome. Most of the decisions made in any organization are the category I decisions and usually happen at lower levels of management
	Category II	Decisions that are nonroutine and nonrecurring, with a lot of uncertainty involved in the outcome. These are primary decisions made mostly by middle and upper level managers

The most important activity of a manager is to make decisions through the process of choosing between two or more alternatives for solving the problem (Bazerman & Moore, 2009; Negulescu, 2014). Uncertainty is a factor that can never be eliminated from management decisions, but its effects and the number of times it occurs can be lessened by using management theories and concepts. In the end, to know if the decision was successful, managers must monitor the situation after the implementation of the decision (Harrison & Pelletier, 2000). Decision-making is the most important managerial activity that has two aspects—the act and the process (Hitt et al., 2014). The aspects of decision-making are creativity, participation, meeting, negotiation, power, and styles of decision-making (Marcus & Dam, 2015). Decisions can be required for a complex issue like manufacturing companies wanting to decide whether to develop multiple production sites for various reasons from cheaper labor to access local markets (Julka, 2008).

4.2.2 *Individual and Group Decision-Making*

Decision-making depends on a specific problem, which has three factors: the character of the problem, which can be routine or specific; the situation, which can be urgent or not urgent; and with those involved, which can be an individual or group (Marcus & Dam, 2015). Decision-making techniques may vary depending on whether it is done individually or by a group (Hitt et al., 2014). Individual decision-making can be made faster and depends greatly on the efficiency, knowledge, and experience of the individual responsible for the decision, while group decision-making can be frustrating and stressful as well as costly and time consuming (Pollard, 1987; Mullins & Christy, 2013). Group decision-making tends to be slower to reach conclusions in comparison to individual decision-making although it has certain advantages in terms of greater cumulative knowledge to problems (Hitt et al., 2014).

Working together can, however, result in a much better outcome than working individually and they can make objectively better decisions. Decision-makers in a group are ready to take more risks than they would have taken it as an individual as Mullins and Christy (2013) described, "A decision which is everyone's is the responsibility of no one" (p. 554). If the example of a decision is taken, such as the cleaning of a river by all plastics to be removed and no longer disposed of in the river, it might involve the regulatory committee, civic organizations, the general public, the government policymakers, etc., and no single member can take the final decision because it is a group decision and each stakeholder have their influence on it (Li et al., 2010). The interaction process between the teams who produce decisions is especially important because each strategic decision denotes a specific combination of the diverse skills, knowledge, abilities, and perspectives. In the interaction processes, the team members identify, retrieve, and integrate their perspectives to decision-making (Amason, 1996). However, for decisions to be made, it is required that there should be disagreements, which can allow the understanding of the repercussions of the decision when it is taken and sometimes the disagreements can provide alternatives (Drucker, 1974).

The Japanese style of decision-making relies on consensus management, which has always worked out favorably for them (Drucker, 1974; Krass, 1997). Having a consensus is of grave importance to finalize a decision; but to implement a decision, however, the consensus in a group must be more than just a simple agreement as a strategic decision requires active cooperation among team members to implement the decision. Understanding the rationale of the decision and commitment to implement the decision both should be included during the consensus in group decision-making (Amason, 1996).

4.2.3 Decision-Making Processes and Models

For a decision to be made, it is required to first identify all possible alternatives and consequences and this situation is known as the condition of perfect knowledge (Roberts, 1999). The decision-maker should be able to rank consistently the value of each outcome, rationally, making it the condition of perfect judgment. It is not possible to achieve both the above conditions and the reason for it is limited knowledge, time, or ability to tackle the situation by the decision-maker (Roberts, 1999).

Rationality on decision-making deals with the assumption that the decision-maker behaves rationally and logically and selects a decision alternative with the greatest value (Roberts, 1999). It is difficult for the decision-maker to be fully rational and thus make the decision appear irrational because the nature of the decisions makes rationality inappropriate. The concept of limited rationality considers all influencing factors with which decisions can be made. A rational decision can be defined using the assumptions of limited rationality by using four principles, which are (Roberts, 1999):

(a) a rational decision is the one chosen from the specified number set of possible decisions,
(b) rational decision depends on decision rule to identify the set of possible decisions,
(c) the rational decision applies to a specific situation and may differ between decision-makers, and
(d) rational decisions depend on the information available.

Managers in organizations believe that there is no place for nonrational and irrational decisions, but they are often put into circumstances where they are surrounded by intuitions and judgments and this problem was solved by efficient computer systems, which aided as decision-making tools (Holloman, 1989). The risk of irrational and nonrational decision-making still exists because human judgments and intuitions are a part of this process in the form of choice of variables, mode of use, and other parameters (Holloman, 1989). Rationality can be about practical rationality on what to do and/or theoretical rationality on what to believe (Walker, 2010).

Practical rationality and theoretical rationality are intricately connected as reasoning about what to do relies in many ways on what one believes (Walker, 2010). In terms of rationality, it was common for researchers to exclude emotions from reasoning understanding that emotions most likely interfered with the rational decision-making process (Li et al., 2014). It is further noted that emotions impact decision-making, which leads to the two types of decision-making: rational and irrational decision-making leading to the understanding that past experiences are linked to the predictions of future in terms of possible outcomes of decision-making (Li et al., 2014). The models and classification of decision-making vary in many studies and some of the more prominent ones are shown in Table 4.2 (Nooraie, 2008; Bazerman & Moore, 2009; Slater, 2003; Rode, 1997; Harrison & Pelletier, 2000; Donnelly et al., 1978).

Table 4.2 Classification and models of decision-making according to various studies

#	Decision-making models and classification	Description
1	The rational model	In rational decision-making, information is an important factor as an input and as an output and it has five phases: defining the problem, devising the alternatives, evaluating the alternatives, making a choice, and implementing and monitoring the decision
	The irrational model	The irrational decision-making is closer to reality since rational decision-making is an ideal model that is hard to achieve. Irrational decision-making can be identified by four types of processes based on the degree of formality, which deals with the issuing of rules; and the degree of centralization, which deals with the top-down influence of decisions
2	The rational model	Derived from the structure of the situation
	The irrational model	Derived from the structure of the person
	The nonrational model	Relies on the making of a decision realizing that one can never know the basis for the decision
3	The rational or classical model	The rational decision-making model involves the formulation and solution phase of decision-making
	The bounded rationality model	The bounded rationality model developed by Nobel prize winner Herbert Simon is one where people are not always completely rational when they make decisions; the reason being that the situations they confront demand greater information processing capabilities than they possess, which makes their decision-making rationale, but limited or bounded
	The retrospective model	The retrospective model, which is based on decision-makers attempts at rationalizing a decision after they are made
4	The light side model	The light side is the rational decision-making that is prescriptive in nature and analytical
	The dark side model	The dark side is the nonrational decision and the irrational decision making and they are descriptive and are intuitive. Nonrational decision-making is decision made by intuition and judgments, and irrational decision-making involves feelings and emotions in interpersonal relationships

Decision-making is the most relevant part of the administration and the results of decisions depend on the way that is used in making the decisions. The bounded rationality theory by Herbert Simon explains the idea that humans are only moderately rational and that none of the decisions in an organization can be the function of any one individual (Kalantari, 2010). The classical and neoclassical theories rely mainly on rationality in terms of gathering the important information about the issues currently under consideration and in which the decision-maker is considered as an observer in the decision-making, not an actor. Simon concluded that the classical and neoclassical approaches in dealing with decision-making concepts are unrealistic and ignore limitations that are involved with gathering the necessary information to decide including, time, cost, organizational culture, etc. in arriving at a decision (Kalantari, 2010). Many studies explain it in various phases or steps as shown in Table 4.3 (Nooraie, 2008; Hitt et al., 2014; Back, 1986; Bazerman & Moore, 2009; Pollard, 1987; Negulescu, 2014; Sabaei et al., 2015; Pownall, 2012; Donnelly et al., 1978).

Rationality has not been considered by many when managerial decision-making is concerned because managers are expected to take optimal decisions but there are rational factors that do not allow them to do so (Rode, 1997). Management decisions are rational and focused on the long-term objectives of an organization; no other type of decision can justify an organization as much as the rational model (Harrison & Pelletier, 2000). Decision-making does not have a fixed procedure and it is a dynamic process (Donnelly et al., 1978). The decision-making process takes into consideration alternatives, risks, and potential outcomes, and then a decision is reached (Mele, 2010). All models of decision-making begin with the process of identifying the problem; it is not possible to make an entirely rational decision due to the emotional elements (Negulescu, 2014). The decision-making process contains various steps before the final decision is made.

4.2.4 Strategic Decision-Making

Top management deals most of the time with strategic decisions that are not well structured or have a proper routine but are important to the firm, in which top management plays a crucial role (Nooraie, 2008). Decision-making by managers is characterized by a high degree of political activity and performed more in the top level of management while in the implementation stage, political activity plays an important role in the hands of the middle-level managers to delay, redirect or sometimes sabotage the implementation process. Decisions that are strategic for an organization are complex but especially important for the function, survival, and development of the organization (Bailey & Johnson, 1991). The process by which strategies are developed is of considerable importance in that strategy must align the organization with its business environment. If the processes by which strategies are made are understood clearly, then the organization will be in a better position of developing and

Table 4.3 Stages of decision-making according to various studies

#	Stages of decision-making	Description of steps
1	Two phases	(a) The decision-making phase in which the quality of the decision-making process outcome can be evaluated, (b) the implementation phase, which determines how well the chosen alternative is accomplished
2	Two-step process	(a) Formulating, which is about identifying a problem or opportunity, getting information about it and diagnosing the factors that affect it, (b) the solution, which requires finding alternatives, selecting a preferred solution, and implementing the desired course of action
3	Four-stage process	(a) Recognizing and defining the need, (b) the search for alternative solutions, (c) evaluation of alternatives, (d) choosing from the alternatives
4	Five-stage process	(a) Define the problem, (b) determine the evaluative criteria, (c) identify all possible solutions, (d) judge which solution works the best for the relevant criteria, and (e) choose the optimal solution
5	Six-step process	(a) Define the problem, (b) identify the criteria, (c) weigh the criteria, (d) generate alternatives for the problem, (e) rate each alternative on each criterion, and (f) come up with the optimal decision
6	Seven-step process	(a) Defining the problem, (b) identifying the limiting factors, (c) developing potential solutions, (d) analyzing the alternatives, (e) choosing the alternative, (f) implementing the decision, and (g) keeping a control and monitoring system
7	Eight-step process	(a) Defining the problem, (b) determining the requirements, (c) establishing goals, (d) identifying alternatives, (e) developing evaluation criteria, (f) selecting decision-making tools, (g) applying the tools, (h) checking the answer
8	Nine-step process	(a) Establishing specific objectives, (b) measuring results, (c) identifying problems, (d) developing alternative solutions, (e) developing payoff tables which leads to certainty, (f) assessing uncertainty or risk conditions, (g) choosing an alternative, (h) implementing the decision, and (i) practicing control and evaluation

implementing effective and efficient processes of strategic decision-making (Bailey & Johnson, 1991).

Strategic decisions are usually nonroutine and not well structured and thus do not have a standard formula that is of special concern for large organizations with enormous resources (Bailey & Johnson, 1991). The process of strategic decision-making can be explained by introducing six perspectives: the planning perspective, the logical incremental perspective, the political perspective, the interpretive perspective, the visionary perspective, and the ecological perspective. The planning perspective considers that the decision-making process will be planned until an optimal solution is achieved without an element of surprise (Bailey & Johnson, 1991).

The logical perspective, unlike the planning perspective, reanalyzes each step and allows the making of changes and tactical realignment of the existing strategies in a logical manner (Bailey & Johnson, 1991). The political perspective is based on the negotiation and bargaining power of the individuals in the interest group and a strategy that is acceptable to the most powerful group will be chosen accordingly. The interpretive perspective considers the organization's aspects in terms of shared beliefs, business environment, past experiences, and the organizational culture as the basis of strategic decision-making.

The visionary perspective is directed by the vision for the future of the organization and is based on intuition and understanding of the organization's strategy. The ecological perspective is based on the natural course of an organization to change in terms of processes, structures, and systems. The six perspectives may show some agreement with each other but are not mutually exclusive (Bailey & Johnson, 1991).

4.3 Efficient Decision-Making Processes

Survival, which includes preservation and expansion of economic, competitive, and social roles and growth are particularly important corporate objectives for every organization (Donaldson & Lorsch, 1983). Decision-making is about selecting choices or making compromises to meet the business objectives and improved decision-making can be the key to better organizational performance (Harvey & Service, 2007). Managers do not make decisions only to increase the shareholders' wealth, but the concern is also with expanding the organization's assets including the human assets (Donaldson & Lorsch, 1983). Two reactions could trigger a decision-making process: a problem needed to be solved and an opportunity that could be used (Bulog, 2016). A manager's business environment plays a crucial role along with the individual characteristics of the manager in effective decision-making (Bridge & Dodds, 2018).

4.3.1 Effective Managerial Decision-Making

Managers affect strategic decisions which in turn affect organizational performance (Bridge & Dodds, 2018). Managerial decision-making depended a lot on rationality and intuitiveness (Donaldson & Lorsch, 1983). While making decisions, managers need to choose a side and when two options are on the table, the one side winning the argument does not consider whether the decision will be effective or solve the problem in hand; the focus here is more on consensus than on the effectiveness of the decision (Drucker, 1974). The most important problem managers face while making decisions is the influence of effective decision-making (Hitt et al., 2014).

However, maximizing behavior is the behavior of the decision-maker to be economic and is based on the hypothesis that the decision-maker is rational (Mele,

2010). The decision maker's choice is based on maximizing their preferences, rationality that relies on an understanding of efficiency to obtain a certain objective looking into the benefits that can be developed cost wise. For decisions to be effective, managers should start with opinions rather than facts, which later should be tested against reality (Drucker, 1974). Three characteristics are important in this case: the decision-makers' characteristics, the characteristic of the problem, and the environment's characteristics (Hitt et al., 2014).

Decisions may end up with good results or bad results, but the top managerial positions have a prime responsibility to make important decisions that need to be effective (Krass, 1997). The top management consisting of managers with different capabilities makes more innovative, good quality decisions than managers with less different capabilities (Amason, 1996). Decision-making at the operational level can be efficient with the use of performance measurement systems; although the use of such systems is limited many times only for evaluations and review of performance after a period. Managers should use performance measurement systems to monitor and control performance and make decisions accordingly (Moreira et al., 2013).

Managerial decision-making means that decisions are neither only economic, relating to efficiency nor exclusively ethical, relating to the greater good of mankind. Managers, therefore, need integral rationality, which would include both instrumental and practical rationality, of which practical rationality should be the main driver for decision-making (Mele, 2010). Industrial decision-makers, traditionally, minimize the cost of expansion when considering factors like manpower availability and logistics connectivity at the beginning, which causes the decision of expansion to fail at the end (Julka, 2008). In 2012, EY conducted a survey that included 285 senior executives globally from the various consumer products sector and 81% of the respondents who participated in the survey agreed that they needed to make swifter decisions and increase their level of insight (EY, 2013). Conventional decision-making focuses more on financial statement items and ad hoc reporting and excludes key external and operational drivers, which limit a company's capability to make fully informed decisions (EY, 2013). The factors that complicate decision-making are (Marcus & Dam, 2015):

- the multiplicity of criteria,
- interdisciplinary input,
- joint decision-making,
- risk and uncertainty,
- long-term consequences,
- value judgments, and
- intangible factors.

Excluding the effect of emotions increases the complexity of decision-making and reduces the effectiveness of the decision (Harrison & Pelletier, 2000). Managerial decision-making is a process through which managers are expected to fulfill the required duties and responsibilities; in response to the outcome of the decision, the managers must be rewarded for decision-making that is effective and criticized for

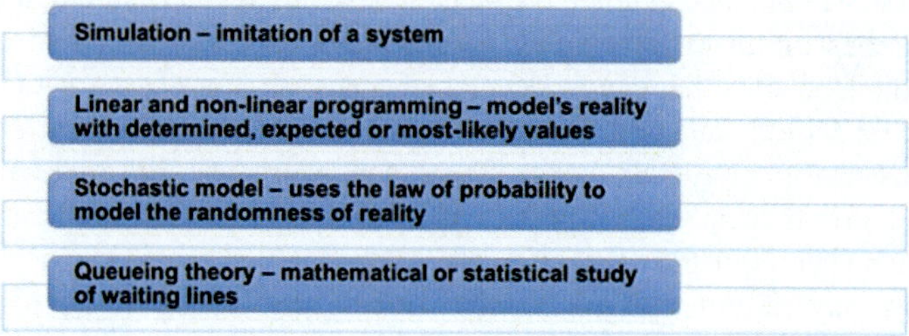

Fig. 4.1 Decision-making using operations research method

decision-making that is a failure. Effective managerial decisions are vital for long-term organizational growth and prosperity (Li et al., 2014). Participatory decision-making (PDM) refers to the decisions made only by the managers and not the subordinates and managers who believe that participatory decision-making is going to enhance the effectiveness of productivity in an organization then the manager is most likely to use it (Parnell & Crandall, 2001).

Organizations have information or data, but sometimes managers do not have access to it, which affects strategic, effective, and timely decisions (EY, 2013). All decision-makers must have the same set of facts, instead of having it in bits and pieces, and this will ensure the success of the business (Slater, 2003). Any piece of information that supports a conclusion is known as evidence and when managers make decisions, they should have supporting evidence that is concrete and critical to the decisions (Denyer, 2013). The questions must be framed properly for the organizations to make the right decisions along with finding the right data to support the analysis and all this must be done in a way that is transparent and repeatable. To make effective decisions, managers today use techniques and tools that aid in helping them take an informed and efficient decision (Lee, Oh, & Pines, 2008).To enhance the decision-making processes in an organization, operations research (OR) as a technique can be used that uses methods as shown in Fig. 4.1 (Lee, Oh, & Pines, 2008).

4.3.2 Tools and Models Used for Efficient Decision-Making

To enhance the capability of decision-making, managers can use various tools and techniques (Roberts, 1999). One such technique is decision analysis, which requires the decision problem to be broken down into a set of smaller and easier problems and each one of the small problems is dealt with separately. The result of each small problem is integrated into the end and this results in the making of the main decision. The most important step in decision analysis is to define the problem for which a solution is required. Defining the problem includes identification of the decision problem and defining the decision criteria, alternatives, and objectives. The key decision group must be chosen keeping in mind all the stakeholders because it will affect the decision criteria, alternatives, and objectives (Roberts, 1999).

Businesses today use various models to increase the efficiency of decision-making. A recent model of decision-making is the Multiple Criteria Decision-Making (MCDM), which combines the performance of decision alternatives against several, contradictory, qualitative, and/or quantitative criteria (Kolios et al., 2016). The MCDM results in solutions that are finalized after considering trade-offs, interconnections, risks, and uncertainties of the criteria and alternatives. The MCDM uses a decision table to help in making a clear and concise decision (Sabaei et al., 2015).

Multiple attribute decision-making (MADM) is another model that usually involves the use of criteria that are subjective and relate also to the preferences of the decision-maker related to their past choices (Bitarafan & Ataei, 2004). Two other approaches to decision-making are the Multi-Attribute Utility Approach (MAUA) and Recognition Primed Decision (RPD). MAUA is a systematic process used by new or inexperienced decision-makers that requires a decision to be justified, while RPD is an approach used by experienced decision-makers for decisions involving dynamic situations, high stakes, and time pressures (Raza, 2009).

The models for decision-making are used by the managers by giving scenarios again and again until the senior management finds a solution like the thought process they have for alternatives for a decision (Rode, 1997). For the decision-making process to be effective and efficient, emphasis should be given to technical competencies along with human relations skills (Itanyi et al., 2012). Traditionally, decision support is regarded as a technical tool that correlates with computer applications that perform this role (Paprika et al., 2008). Decision-making techniques like balanced scorecard, decision matrix, and decision trees, as well as systems for decision-supporting like expert systems and simulation, help in accelerating the decision-making processes (Marcus & Dam, 2015).

The reason for decision-making information systems to be used frequently is that large amounts of information must be analyzed, decisions made need to be swifter, and good decisions need to be made, so sophisticated systems and information security are paramount to the process (Baltzan & Welsh, 2015). In the case of another technique named fuzzy logic, multiple attribute decision-making can be performed by selecting an alternative from a set of alternatives that are categorized in terms of different attributes, the technique involves the observance of a single objective. The selection of an alternative is made based on a score determined from the levels of attributes of each alternative and the classification of alternatives is done using a role model defined based on the same case outcomes (Bitarafan & Ataei, 2004).

4.3.3 *Techniques Used for Efficient Group Decision-Making*

Organizational decisions are also made in a group, those who are involved in the decision-making process form the decision body and the decision body should be dynamic (Pownall, 2012). Although human input is the main factor in decision-making, the nonhuman factors constituting relevant data, information, and knowledge, cannot be neglected and form the decision context. The key stages in the rational model of decision-making are problem definition, determination of criteria for evaluation, identification of all possible solutions, judging the outcome about preferred criteria and aim, and choosing the most optimal solution (Pownall, 2012). There is no consensus when trying to understand whether the group decision-making is more effective than individual decision-making. The attribution theory plays an important role in the effectiveness of decisions made in a group. It deals with the requirement of individuals in a group to take credit for the success and attributing failure to external circumstances (Pownall, 2012). The methods to make effective group decisions are described by Pownall as illustrated in Fig. 4.2.

Brainstorming deals with the verbal interaction between group members to express free thinking about ideas while brainwriting is a silent method where group members write down their ideas and all ideas are further discussed to further form new ideas (Pownall, 2012). A combination of brainstorming and brainwriting is known as the nominal group technique. Buzz sessions are group discussions, the only difference being that a single group is divided into many subgroups for discussions. Quality circles involve employees in the decision-making process to understand and make decisions to operational concerns (Pownall, 2012).

Decision-making and problem-solving are related but are not interchangeable terms (Pownall, 2012). Decisions do not fail because a theory like the normative theory is applied; they fail because the decision-maker, due to his lack of knowledge, is incapable of making a better decision. The information may sometimes be enough, but there is less clarification and a typical Management Information System (MIS) ignores the impact of bounded rationality in managerial decision making (Rode, 1997). Evidence-based management requires the manager to do the following during a decision-making process: practitioner expertise, local evidence, existing research evidence, opinions, choice and ethics, and in the end, mapping of all the evidence collected through the above (Denyer, 2013).

The decision-makers should not start with the process of decision-making by assuming one solution is right while others are wrong allowing bias in considering other alternatives (Drucker, 1974). One of the alternatives should always be to not

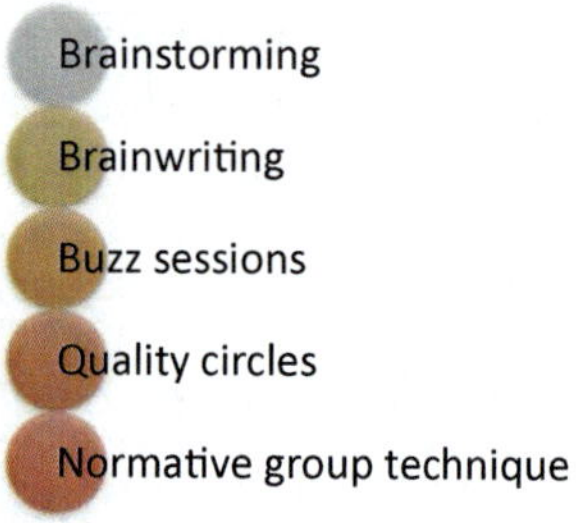

Fig. 4.2 Methods for effective group decision-making

do anything about the situation, which lets the decision-maker thinks if the decision-making is necessary based on cost and risk factors. Decisions taken need not be always right and they can go wrong due to various reasons: choices that were not considered, information on the options taken might not have been right, knowledge of the existing circumstance not being right, the experience of the past may be irrelevant, and the wrong prediction of the future, are some of the causes for decisions going wrong (Heller, 1989). Drucker explained that there is no perfect decision, one must always pay a price for the decision taken.

4.4 Decision-Making in LSOPM Operations

Decision-making in large-scale open-pit mining operations requires an understanding of various factors of which only some have been addressed in various studies. Mining is the process of removal of material that occurs naturally on earth's surface or subsurface and involves decisions on how to efficiently recover and treat the removed material which can be metallic ores such as gold and copper, or nonmetallic minerals such as sand and gravel, or fossil fuels such as natural gas or petrol (Espinoza et al., 2012).

Decisions regarding starting-up an operation depend on top management. In many countries, management is quite aggressive, and they decide to start the project based on financial modeling even if the LOM is 4 or 5 years. There are many organizations where management waits till they get positive LOM of 10 years or more. In this case, although it depends on many other factors, the preproduction costs continue while they do not make any revenue, which is the reason for many mine operations to not start for many years and changing management due to insecurity of getting the expected rate of returns. Decision-making of continuing preproduction operations or selling the property is a crucial decision for the management of any mining organization.

In the beginning, mainly high-grade ore deposits, which were easily found and exploited as they were exposed at the surface was mined (Dold, 2008). The material can be classified as waste or ore in a mine site by looking at the ratio of the weight of metal or mineral value recovered in the concentrate to 100% of the same constituents of the heads or feed to the process expressed as a percentage (Carrasco et al., 2008). For example, there are very rich copper mines (1–2 wt.% Cu), which have a cut-off ore grade of around 0.6 wt.% Cu (i.e., material with less of 0.6 wt.% is not considered as ore), while other less rich porphyry copper mines operate very successfully with a total ore grade of 0.6–0.7 wt.% Cu (Dold, 2008) (Dold, 2008). Exploration for discoveries depends on the target cut-off ore grade and thus, the decision if a certain material is classified as waste material or ore depends on various factors that include metal price during mining, ore grade of the deposit, personal decisions of the management, a technique available, geo-metallurgical behavior of the ore, the complexity of the ore mineralogy and recovery, etc. (Dold, 2008).

The equipment selection process in LSOPM operations is extremely important and often the final decision is made by the top management (Roman, 1999). However, many people are involved in the decision-making process, such as the operations department and the procurement department. This shows that the decision of equipment selection is handled in both the strategic and operational levels of management. The operations department must be included since the day-to-day operations are carried out using the equipment, but the department that is usually not involved in the decision-making process is the maintenance department (Roman, 1999).

Case Study Outcomes

An informed decision on selecting and procuring equipment requires the maintenance department to be involved in the process since maintenance costs and equipment availability are the primary responsibility of the maintenance department. Theories that match the result of the research on decision-making in LSOPM operations in comparison to general decision-making are given in Fig. 4.3.

To understand the advantages of better decision-making process, it is required to understand the components of decision-making and the use of faster decision-making. In LSOPM, there are decisions to be taken always and according to the outcome of the primary research, taking a decision is extremely important. Not taking a decision is considered worse than taking a wrong decision. Some of the respondents, however, also expressed that you can take the time to take decisions depending on the situation. Participants also responded that effective decision-making is important and the decisions that are taken must be well evaluated. The right information and simplified processes are particularly important for decision-making in LSOPM operations.

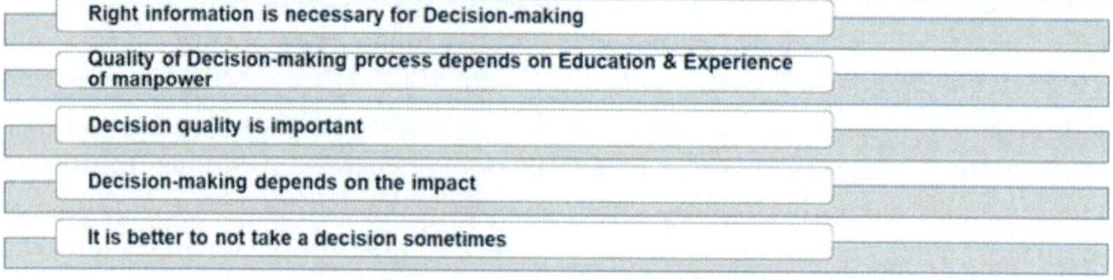

Fig. 4.3 Conventional decision-making concepts applied to LSOPM operations

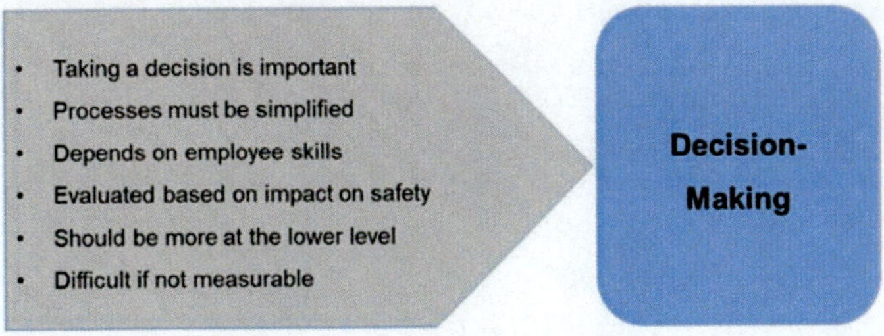

Fig. 4.4 Decision-making concepts applied to LSOPM operations

People knowledgeable of the processes would be able to make the right decisions but the accountability and responsibility for each decision should be made clear. It is also determined that the quality of the decision depends on the education and experience of the people. Decision-making also depends on the availability of data, as not understanding the data and exact benefit of a decision makes it harder to make the decision. The quality of the decision and the impact of a certain decision on any area of the organization are also important. It is also evident from the research that many times, Operational decision-making should be more at the lowest level. New concepts that are the outcome of the research and are not based on other studies are given in Fig. 4.4.

4.4.1 Faster Decision-Making

Decision-making can be faster if the information is available faster and for information to be available faster the LSOPM operations needs to have information technologies. The literature does not stress on faster decision-making in LSOPM operations, however, according to Hitt et al. (2014), faster decision-making is important for any organization and is related to individual decision-making and group decision-making. The decisions are made faster by the former than the latter. In the open-pit mining environment, faster decision-making is important and a decision like where should the equipment be deployed, what is the cycle time and how can it be reduced, are decisions that are taken daily, and require personnel to take faster decisions to not have adverse effects on production. While information technologies do give the information faster, it is also required that the data be analyzed by the right people with the knowledge and skills to be able to take the right decision then.

On the downside, the faster decision-making process can sometimes affect the quality of decisions which means in certain instances taking time to analyze decision-making is also advisable. For example, in a running mine operation, there is an environmental spill with minimum impact but for any quantity of spill even if it is extremely little, it is mandatory to be reported to the government, as per the mining license agreement. The delay in response or deciding by the concerned personnel

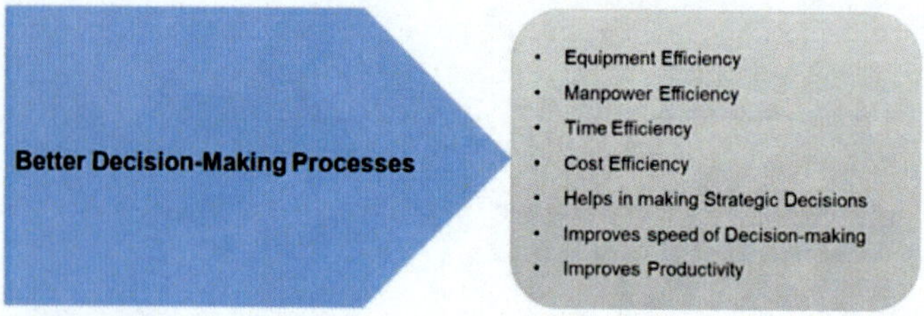

Fig. 4.5 Advantages of better decision-making processes

to report the government can cost the company the danger of getting their mining license revoked.

4.4.2 *Advantages of Better Decision-Making Processes*

As in any other industry, a better decision-making process also has its advantages in LSOPM operations. The advantages are in the form of achieving efficiency in equipment, manpower resources, time, cost, and the overall efficiency of the operations. In this case, a better decision-making process ensures faster decision-making and can also help in negotiations in the upper level of management. One of the benefits is related to making good strategic decisions as decision-making at the top level is extremely important for every organization. In today's borderless business world, strategic decisions need to be made beyond geographical boundaries and a better decision-making process will enhance the strategic decisions made at top-level management.

Productivity and profitability of LSOPM operations can also be increased if the decision-making processes are better. It enables us to do all the required planning in terms of mine design, the requirement of equipment, the requirement of manpower, and other resources, beforehand. It also makes the short mine communications systems more effective. The benefits of better decision-making processes are given in Fig. 4.5. A list of all possible outcomes for the benefits of better decision-making process in LSOPM operations is given in Table 4.4 in the appendix section at the end of the chapter.

4.4.3 *The Interrelationship Between Swift Decision-Making and Productivity*

Swift decision-making in LSOPM operations has a positive effect on productivity in terms of efficiencies. Decision-making can be operational or strategic. Time is a very important factor in decision-making, a delay in a decision on whether a piece

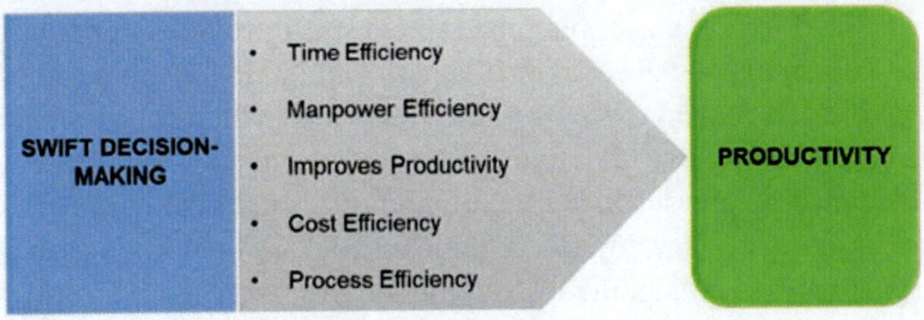

Fig. 4.6 Effect of swift decision-making of productivity

of equipment should be operated even after a technical problem has been detected or should it be left idle, could end up resulting in major problems in the equipment or lesser production. Both the consequences depend on how fast the decision is made to stop or to continue with the operations of the equipment. This may not be the case for all types of decisions so sometimes it depends on the situation.

The above decision was an operational decision. In strategic decision-making, decisions are made on a longer term basis, which directly impacts future Operations, Life of the Operations, Cash flow, Capital investments, Corporate Social responsibility, Environmental impacts, etc. For example, whether to start production for a new operation when the market price of the mineral is low or to wait for the market price to increase, this has a bearing on time for startup or expansion impacts costs, resulting in impacting in cost efficiency of the Operations.

In an LSOPM operation if manpower utilization is not done efficiently it can affect productivity. Low metal prices would require the optimization of various resources. One such resource was manpower resources, and optimization of manpower resources can sometimes also mean downsizing. If the measures of optimization require downsizing then the decision must be faster, a slow decision would increase costs and decrease in revenue and profits, ultimately resulting in the danger of closure of operations. A summary of the main findings of the effect of swift decision-making on productivity is given in Fig. 4.6.

Decision-making in terms of which equipment should be part of which fleet in an area of production is crucial and this requires continuous real-time information provided by information technologies. Using information technology minimizes risks when it helps in having a whole distribution of likelihoods in some scenarios, which helps in making less risky decisions. Information systems help in using strategic opportunities by maximizing the value of the operation and instead of assisting in better decision-making it helps in better integration of data. It helps in streamlining the processes and operations like a better use of the warehouse, automatically reorder parts, which are below the minimum level, to know what your optimal minimum and maximum levels are, plan for maintenance of when your engines need to be replaced, etc.

Information technologies, like the expert systems, make a lot of information available and give a clear picture of what the snapshot performances of equipment are where the processes are driving decision-making processes. The expert systems assist in the process plant in two ways, first, the system itself provides trending information so the plant personnel can see how it is performing, and second, it replaces the conventional role of a control room operator. Historically, the control room operator monitors the mill and gets lots of outputs about equipment and then makes decisions on factors like varying pressures, if it is too high that could cut hard rock or when power is too high then it is going to cut the hard rock or the mill is making noise, slow the mill down which is going to consequently impact the hard rock, etc. This job of decision-making can now be done by expert systems since it can read more information at a time than an operator and come up to a conclusion faster.

Information technologies can help to stimulate the whole mine area aiding mining personnel in decision-making on which areas are to be mined at what point of time. The simulation assists in many practical scenarios like for example, the process plant has some maintenance work and is not able to process ore for a few days, the personnel can decide based on the simulation to mine in areas where there is more waste and when the plant starts working, the personnel can with the help of the simulation mine as per plan in the areas where it was decided to mine earlier. Many mining organizations do not use the information technologies to its potential or have basic warehouse systems in a legacy state not using simple systems like barcoding for issue and receipt of goods. The flow of information from the pit personnel to the procurement personnel should be automated for efficient operations and better process of decision-making.

The impact of swift decision-making in process efficiency can be explained in terms of the purchase of mining equipment and delivery from a foreign location. There are processes to be followed in overseas transportation and negotiations. Decision-making is crucial in this case and if not taken faster, it may affect process efficiency negatively. When the production process in a mining operation or mill is monitored, if swift decisions are taken by personnel at all levels for any inefficiency, the production can be improved in an LSOPM operation. When production volumes and productivity in a mining operation increases, profitability also increases. The performance of various resources also plays a particularly important role in LSOPM operations. When decision-making is faster in terms of the resources being used, it increases the performance of the individual resources and in terms of returns, for example, if you can make those decisions more quickly, then you can mine more quickly and making more tons per day per hour, that is higher profit. A list of all possible outcomes for the effects of swift decision-making on productivity in LSOPM operations is given in Table 4.5 in the appendix section at the end of the chapter.

4.5 Conclusion

Information is an especially important factor in LSOPM operations and availability of the information which is correct and faster will affect the productivity of a mining operation. It is not enough to just implement information technologies; it should be used efficiently for gathering and analysis of data to information that can be shared.

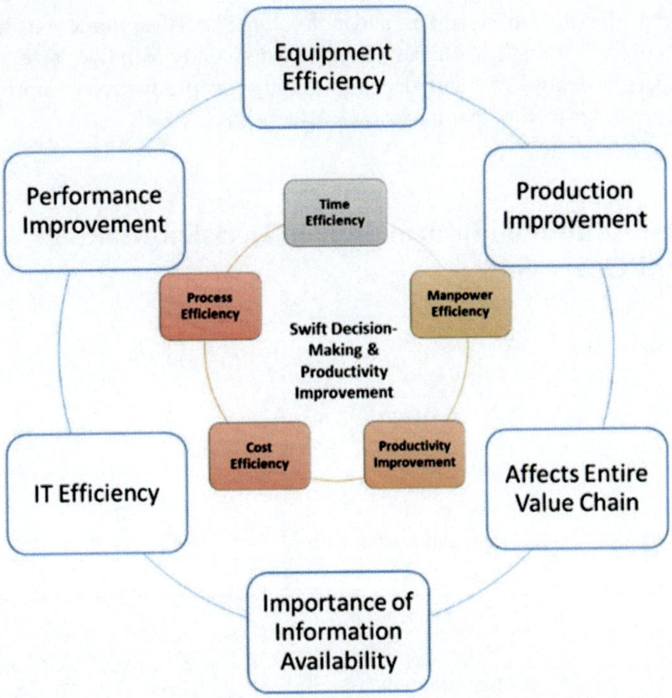

Fig. 4.7 Factors relevant to swift decision-making and productivity improvement

When decision-making is slower, it affects productivity and the economic viability of an LSOPM project.

Improvement in production, efficiency in time, efficiency in equipment utilization, efficiency in the use of manpower resources, cost optimization, increased performance, and profitability all lead to an improvement in productivity as shown in Fig. 4.7. The inner circle includes the factors that are most relevant toward the effects of swift decision-making on productivity and the outer circle depicts the factors which have a lesser effect on productivity in comparison to the inner circle factors.

4.6 Summary

Every aspect in business requires decision-making and it is something that needs to be taken at each level of management in terms of strategic decisions, tactical decisions, and operational decisions. In general, decisions can be classified into different types and models based on the situation and level of management. It is evident from the research that LSOPM operations require faster decision-making and a better decision-making process assists in faster decision-making. Advantages of faster decision-making are enormous, and it directly affects the productivity improvements in an organization. The chapter, in conclusion, ends with five factors that are closely affected because of swift decision-making on productivity improvements along with six other factors that have some effects on it.

Appendix: Additional Information on Decision-Making in LSOPM Operations

See Tables 4.4 and 4.5

Table 4.4 Benefits of better decision-making process

#	Description
1	Time efficiency
2	Manpower efficiency
3	Improves productivity
4	Cost efficiency
5	Process efficiency
6	Equipment efficiency
7	Improves production
8	Affects entire value chain
9	Information availability is important
10	Slow DM affects productivity
11	Affects project viability
12	Not directly related
13	Slow decisions cause longer lead times
14	IT efficiency is important
15	Improves performance
16	Depends on the situation
17	Improves profitability

Table 4.5 Effect of swift decision-making on productivity

#	Description
1	Taking a decision is important
2	Decisions need to be taken based on available information
3	Decision-making should be well evaluated
4	Effective decision-making is important
5	Simplified processes are important
6	Impact of decision on safety
7	Accountability and responsibility should be clear
8	People close to process take right decisions
9	Right information is necessary for Decision-making
10	Quality of decision-making process depends on education and experience of manpower
11	Not taking decisions is worse than taking wrong decisions
12	Decision-making is easy when you have the numbers
13	Decision-making is difficult when you cannot see the exact benefit
14	Decision quality is important
15	Decision-making depends on the impact
16	Depends on employee skills
17	Decision-making should be more at the lowest level

References

Amason, A. C. (1996, February). Distinguishing the Effects of Functional and Dysfunctional Conflict on Strategic Decision Making: Resolving a Paradox for Top Management Teams. *The Academy of Management Journal, 39*(1), 123–148. Retrieved from https://www.jstor.org/stable/256633.

Anthony, N. R., & Govindarajan, V. (2007). The nature of management control systems. In N. R. Anthony & V. Govindarajan (Eds.), *Management control systems* (12th ed., pp. 1–17). The McGraw-Hill Companies.

Anthony, N. R., Govindarajan, V., Hartmann, G. H., Kraus, K., & Nilsson, G. (2014). Designing management control systems. In N. R. Anthony, V. Govindarajan, G. H. Hartmann, K. Kraus, & G. Nilsson (Eds.), *Management control systems* (pp. 34–62). Berkshire: McGraw-Hill Education.

Back, W. K. (1986). Decisions under uncertainty: Rational, irrational, and non-rational. *The American Behavioral Scientist*, 14–19.

Bailey, A., & Johnson, G. (1991). *Perspectives of the process of strategic decision-making*. Cranfield University.

Baltzan, P., & Welsh, C. (2015). Decision making and business processes. In P. Baltzan, & C. Welsh (Eds.), *Business driven information systems* (pp. 30–62). McGraw Hill Education.

Bazerman, H. M., & Moore, A. D. (2009). *Judgment in Managerial Decision Making* (7th ed.). New Jersey: John Wiley & Sons Inc.

Bitarafan, M. R., & Ataei, M. (2004, October). Mining method selection by multiple criteria decision making tools. *The Journal of the South African Institute of Mining and Metallurgy*, 493–498.

Bridge, J., & Dodds, J. C. (2018). The firm and managerial decisions. In J. Bridge, & J. C. Dodds, *Managerial Decision-Making* (pp. 1–23). New York: Routledge: Taylor & Francis Group.

Bulog, I. (2016, July 6). The influence of top management demographic characteristics on decision-making approaches. *ECONVIEWS*, 393–403.

Carrasco, P., Chiles, J.-P., & Seguret, S. (2008). *Additivity, metallurgical recovery, and grade*. Santiago: GEOSTATS.

Denyer, D. (2013). Are you an evidence-based manager? *Management Focus*, 24–25.

Dold, B. (2008, September 27). Sustainability in metal mining: From exploration, over-processing to mine waste management.*Reviews in Environmental Science and Bio/Technology*, 275–285. doi:https://doi.org/10.1007/s11157-008-9142-y.

Donaldson, G., & Lorsch, W. J. (1983). *Decision making at the top*. New York: Basic Books Inc., Publishers.

Donnelly, H. J., Gibson, L. J., & Ivancevich, M. J. (1978). *Fundamentals of management: Functions, behavior, models*. Dallas: Business Publications Inc.

Drucker, F. P. (1974). *Management: Tasks, responsibilities, practices*. New York: Harper & Row, Publishers Inc.

Espinoza, D., Goycoolea, M., Moreno, E., & Newman, A. (2012, December 6). MineLib: A library of open pit mining problems. *Annals of Operations Research*, 93–114.

EY. . (2013). The future of decision making. *Insights for Executives, 5,* 1–8.

Fiscor, S. (2007). Australian mining technology. *Engineering and Mining Journal, 208*(6), 41–49.

Haider, S., & Mariotti, F. (2016). Unfolding critical events and strategic decisions: The role of spatial and temporal cognition. *Management Decision, 54*(7), 1813–1842. https://doi.org/10.1108/MD-11-2015-0521

Harrison, E. F., & Pelletier, M. (2000). The essence of management decision. *Management Decision*, 462–469. Retrieved from https://search.proquest.com/business/docview/212061917/CE0896B24DB94C52PQ/20?accountid=150425.

Harvey, J., & Service, T. I. (2007). *Effective decision making*. London: The Chartered Institute of Management Accountants.

Heller, R. (1989). *The decision makers: The people and methods of great achievement*. London: Hodder and Stoughton.

HIA Ltd. (2015). *Vegetable industry development program: Business decision making*. Horticulture Innovation Australia Limited.

Hicks, G. H., & Gulliett, R. C. (1981). *Management*. New York: McGraw Hill Inc.

Hitt, M., Black, S., & Porter, W. L. (2014). Individual and group decision making. In M. Hitt, S. Black, & W. L. Porter (Eds.), *Management* (3rd ed., pp. 325–350). Edinburgh Gate: Pearson education limited.

Holloman, R. (1989). The Light And Dark Sides of Managerial Decision-Making. *Industrial Management*, 15–18.

Itanyi, O., Ewurum, F. J., & Ukpere, I. W. (2012, November 7). Evaluation of decision making criteria with special reference to quantitative and qualitative paradigms. *African Journal of Business Management, 6*(44), 11110–11117.

Julka, N. (2008). *An advanced decision process for capacity expansion in manufacturing networks*. Cranfield University.

Kalantari, B. (2010). Herbert A. Simon on making decisions: Enduring insights and bounded rationality. *Journal of Management History, 16*(4), 509–520.

Kolios, A., Mytilinou, V., Lozano-Minguez, E., & Salonitis, K. (2016, July 21). A comparative study of multiple-criteria decision-making methods under stochastic inputs. (F. Blaabjerg, Ed.) *Energies 2016*. doi:https://doi.org/10.3390/en9070566.

Krass, P. (1997). *The book of business wisdom*. New York: John Wiley & Sons Inc.

Lee, E., Oh, J.-Y., & Pines, E. (2008). Practical managerial decision making tools: Operations research. *The Journal of Applied Business and Economics*, 11–18.

Li, J., Kendall, G., Pollard, S., Soane, E., Davies, G., & Bai, R. (2010). A decision support approach for group decision making under risk and uncertainty. *Internation conference on logistics systems and intelligent management* (pp. 1856–1860). Harbin Heilongjiang: IEEE.

Li, Y., Ashkanasy, N. M., & Ahlstrom, D. (2014, March). The rationality of emotions: A hybrid process model of decision-making under uncertainty. *Asia Pacific Journal of Management*, 293–308.

Marcus, J., & Dam, v. N. (2015). Decision-making. In J. Marcus, & v. N. Dam (Eds.), *Organisation & management: An international approach* (3rd ed., pp. 345–384). Houten: Noordhoff Uitgevers bv.

Meagher, K., & Wait, A. (2010). *Decision making within organizations.*

Mele, D. (2010). Practical wisdom in managerial decision making. *The Journal of Management Development, 29*(7), 637–645. https://doi.org/10.1108/02621711011059068

Moreira, M., Tjahjono, B., & Juliao, J. (2013). Applying performance measures to support informed decision making at an operational level. *The 11th International Conference on Manufacturing Research (ICMR2013), Cranfield University*, (pp. 563–568).

Mullins, J. L., & Christy, G. (2013). Working in groups and teams. In J. L. Mullins & G. Christy (Eds.), *Management & organisational behavior* (10th ed., pp. 332–363). Edinburgh Gate: Pearson Education Limited.

Natale, S. M., Libertella, A. F., & Rothschild, B. M. (1995). Decision-making process: The key to quality decisions. *American Journal of Management Development*, 5–8.

Negulescu, O.-H. (2014). Using a decision-making process model in strategic management. *Review of General Management, 19*(1), 111–123.

Nooraie, M. (2008). Decision magnitude of impact and strategic decision-making process output: The mediating impact of the rationality of the decision-making process. *Management Decision*, 640–655.

Paprika, Z. Z., Wimmer, A., & Szanto, R. (2008). Managerial decision making and competitiveness: The case of hungary. *Competitiveness Review: An International Business Journal*, 154–167.

Parnell, J. A., & Crandall, W. (2001). Rethinking participative decision making: A refinement of the propensity for participative decision making scale. *Personnel Review*, 523–535.

Pollard, W. E. (1987). Decision making and the use of evaluation research. *The American Behavioral Scientist*, 661–676.

Pownall, I. (2012). Effective management decision making: an introduction. Retrieved from www.bookboon.com.

Raza, M. (2009). *Command agents with human-like decision making strategies.* Cranfield University.

Roberts, M. C. (1999). *Risk-based decision-making for the management of structural assets.* Cranfield University.

Rode, D. (1997). *Managerial decision making: Normative and descriptive interactions.*

Roman, P. A. (1999). *Joint enhancement of mine operations and maintenance: A structured analysis approach.* Queen's University (Canada). Ann Arbor: ProQuest, UMI Dissertations Publishing. Retrieved from https://search.proquest.com/business/docview/304542219?accountid=150425.

Sabaei, D., Erkoyuncu, J. A., & Roy, R. (2015). A review of multi-criteria decision making methods for enhanced maintenance delivery. *Procedia CIRP*, 30–35.

Singh, V. P. (2012, June). Managerial decision making. *International Journal of Innovative Research in Commerce & Management, 5*, 1–4.

Slater, R. (2003). The leadership module. In R. Slater (Ed.), *The GE way fieldbook* (pp. 13–39). New Delhi: Tata McGraw-Hill Publishing Company Limited.

Sopta, M., & Slavica, A. (2017). Importance of cost function in business decision-making. *19th International Scientific Conference on Economic and Social Development* (pp. 208–225). Melbourne: 19th International Scientific Conference on Economic and Social Development.

Van Niekerk, G. (2013). From volume to value: Cost optimization in the mining sector. pp. 38–41. Retrieved from https://www.kpmg.com/Global/en/IssuesAndInsights/ArticlesPublications/mining-asset-lifecycle/Documents/from-volume-to-value-v2.pdf.

Walker, L. R. (2010, December). Respect for rational autonomy. *Kennedy Institute of Ethics Journal, 19*(4), 339–366.

Chapter 5
Conclusions and Future Research Direction in LSOPM

Abstract Operational excellence is crucial for every business to sustain in the long run. It is a management philosophy that requires organizations to indulge in introspections, take corrective measures, and focus on continuous improvement. Achieving operational excellence requires the resources of an organization to be used with optimum efficiency. This chapter is in conclusion to the results of the previous four chapters and details the interrelationships between information technology, decision-making, and operational efficiencies. A new model is presented that showcases 13 factors to be considered in LSOPM operations and in general by all large businesses to sustain in the long run. The chapter also includes future recommendations and areas of research in the concerned area.

Keywords Operational efficiencies · Productivity improvements · Decision-making in mining · Information technology in mining

5.1 Introduction

The history of mining reveals that the mining scenario was mainly a stream that concentrated on the extraction of minerals in a higher volume and the productivity factor and optimization of cost were not significant. Operational excellence requires companies to achieve long-term change by going beyond traditional models and initiatives of improvement (Australian Mining, 2016). According to the reports from Deloitte, mining organizations now focus on operational excellence in the form of cutting costs and improving productivity (Australian Mining, 2016). The decrease in demand and the increase in commodity prices have greatly contributed to mining companies to realize that cutting costs can help them survive the crisis (PwC, 2016). However, the high demand for minerals and huge investments in the mining sector requires the focus on defining the concepts of productivity and cost optimization from the industry experts who work hands-on in the large-scale open-pit mining (LSOPM) companies (ICMM, 2012a; Lamghari et al., 2013).

H. Qudrat-Ullah et al., *Operational Sustainability in the Mining Industry*,
Asset Analytics, https://doi.org/10.1007/978-981-15-9027-6_5

In an LSOPM operation, according to each of the chapter, there were key takeaways which are given in Table A5-A of the additional information section at the end of the chapter. From the previous chapters, for LSOPM operations the following are especially important:

(i) Understanding the concepts of cost optimization and productivity due to the variation of demand for metals,
(ii) Understanding the interrelationship between information technology and productivity and information technology and decision-making,
(iii) Understanding the interrelationship between productivity and cost optimization and productivity and equipment utilization,
(iv) Understanding the benefits of better decision-making processes along with the interrelationship between decision-making and productivity.

This chapter is further divided into three sections:

- Interrelationships between organizational efficiencies, information technologies, and better decision-making
- Theoretical contribution to operational excellence in LSOPM operations
- Future research

5.2 Interrelationships Between Organizational Efficiencies, Information Technologies, and Better Decision-Making

It is evident from Chaps. 2 to 4 that organizational efficiencies, information technologies, and better decision-making play a significant role in contributing toward the growth and long-term sustainability of LSOPM operations. The interrelationship between the three factors and the concepts within each also shows that they are very closely related to each other. The interrelationship of each factor with another is further described below.

5.2.1 *Concepts of Productivity and Cost Optimization*

In the mining scenario, production volumes were particularly important in the past. Although there has been a sharp cut in demand for metals, a report from Deloitte (2015) shows that the increase in demand has now begun, especially for gold. Iron ore has also picked up its demand although the market predictions were not positive from 2016. The decrease in demand and the increase in commodity prices have greatly contributed to mining companies to realize that cutting costs can help them survive the crisis. This gravely represents the fact that presently, mining companies need to stress on productivity improvements and optimization of cost. Innovation should be incremental and requires a rethinking of using the equipment, technology, or processes the company already has, in a better way. High demand for minerals and huge investments in the mining sector requires the focus on defining the concepts of productivity and cost optimization from the industry experts who work hands-on in the large-scale open-pit mining companies.

There are costs involved in everything that is done and sometimes it is not possible to directly assign benefits or paybacks in terms of benefits. It is an increase in efficiency that cannot be calculated to clear the payback period. Management must decide whether to go for it or not because the benefits can be seen clearly but numerical values cannot be added to the benefits all the time.

The challenges in mining can vary from one geographical location to another. In one place, most of the challenge comes because of heavy rainfall and annual precipitation while in another place the reason is low temperature or high temperature, or sometimes type of rock-like hard rock, in which case you need to increase your cost in terms of blasting because it will consume more power, each of these factors can contribute to different costs and have to be considered separately to improve productivity.

The market price of the mineral plays a particularly important role in cost optimization. Although strategies for cost optimization should always be used, sometimes it is possible to oversee some costs if the market price of a mineral is extremely high and the overall costs are considerably low to keep operations to continue.

5.2.2 *Interrelationship Between Cost Optimization and Productivity*

Mining companies try to achieve operational excellence by maintaining high productivity and optimized costs, which means striving for sustainable improvement of key performance indicators by applying different types of principles, systems, and tools (Mining Global, 2015). Cost optimization in the mining area should be done in a manner that affects productivity positively. For example, if an equipment's output per day is 50,000 tonnes a day and if for any reason it reduces to 25,000 tonnes on a given day, the cost immediately rises to 150%. The cost of manpower, fuel consumption, etc., will go up meaning the resources used are the same but the output is reduced by half decreasing the productivity and increasing the cost. Another production area-related example is when an excavator produces 2,000 tonnes of material per hour, fixed costs will remain the same, which include the hourly rate of the operator, hourly fuel consumption, maintenance cost, etc. However, if production can be increased from 2,000 to 2,200 tonnes per hour, keeping all other resources the same, the cost will be reduced by 10%. This is the influence of productivity over cost.

The top 40 large-scale open-pit mines, which include coal, oil, and gas mines, have made tremendous efforts and have been successful in implementing cost-cutting initiatives according to the latest surveys in 2015 (PwC, 2016). Two important examples of the initiatives and successes are (a) decrease in cost 2015 by BHP Billiton, a decrease in operating cash costs of $2.7 billion and the generation of volume efficiencies of $1.2 billion due to productivity improvements and (b) mining giant Rio Tinto reported $2.9 billion in cost savings in 2015 and announced that it would further reduce cash costs by $1 billion annually in 2016 and 2017 (PwC, 2016).

However, mining giants like BHP and Rio Tinto made quick cost savings in 2015 by cutting exploration budgets and BHP announced that the company saved $142 million because of the cost-cutting measure in 2015 alone. The top 40 mining companies had a decrease of 17% in operating costs in 2015 with higher production volumes and lower input costs (PwC, 2016). The effect of cost-cutting in terms of productivity initiatives and technological advances like automation is commendable but it is also important to see is if these mining companies can sustain the initiatives for the long term.

5.2.3 Interrelationship Between Equipment Utilization and Productivity

Operational excellence requires companies to achieve long-term change by going beyond traditional models and initiatives of improvement (Mining Global, 2015). Mining equipment like large haul trucks is also sometimes referred to as mobile fleets that continuously move material throughout the mine: queuing, loading, dumping, and hauling both in full and empty modes. In open-pit mining, truck haulage is the largest item in the operating budget, constituting about 50–60% of the total mining costs; it is critical to close the gap and maximize potential truck/shovel capacity (Hanson, 2016). A Caterpillar 777F model haul truck with a nominal payload of 100 tonnes used at one of the open pits in Rosebel Gold Mines, Suriname, is given in Fig. 5.1.

Equipment efficiency does not deal only with optimization of equipment utilization individually, but it requires the optimization of fleets of equipment for a certain area of production. The hauling process in mine operations is complex due to many factors such as weather conditions, processes like blasting, the minerals being mined, human behavior, and uncertainties like mechanical failures, which lead to inefficiencies (Hanson, 2016). OEM is now relentlessly working on increasing the operational efficiency of equipment and decommissioning of the equipment is also done to recycle and use it again (PwC, 2017). The selection of equipment for mining is an especially important decision and is not a standard process as it differs from mine to mine because it involves the interaction of several subjective factors or criteria and complicated decisions (Bazzazi et al., 2009). Equipment in the mining companies has equipment that monitors the machine and the performance of the operator, the knowledge of which is crucial, and many companies try to converge this knowledge to the mine-wide enterprise systems (E&MJ, 2009).

In open-pit mining, the truck-shovel costs are extremely high usually and if it is hard rock mining of gold and base metals then crushing and grinding are the

Fig. 5.1 Haul truck used in LSOPM operations

most expensive (KPMG, 2016). The truck cycle time can be planned well if truck dispatch databases provide the data required for the process and dispatch system. Mining personnel who make crucial decisions should get real-time data and these data can be analyzed to reduce costs, raise productivity, and enhance revenue. The performance management indicators should be realized by companies and should be embedded into business intelligence (KPMG, 2016). It is evident that open-pit mining equipment has an impact on operating costs and is closely related to production volume. The important factors of equipment selection, equipment capacity, and equipment maintenance play a very crucial role in open-pit mining.

5.2.4 Interrelationship Between Decision-Making and Productivity

Operational excellence can be defined as a means for organizations to achieve sustained profitability, aligned with business objectives, and a relevant performance measurement system among other things (Barr & Cook, 2009). Sustainability is possible if there is continuous improvement in terms of processes in a mining organization. A real-time performance measurement system can monitor the improvements taking place. It assists in better decisions by mining personnel where performance is improved by 3–5% enabling operations to earn payback in a short period. The assets of a mining company need to be used in an optimal way to sustain economic viability under continually tough economic pressure (Barr & Cook, 2009).

5.2.5 Interrelationship Between Decision-Making and Information Technologies

The mining industry is quite different from any other industry and its challenges vary in terms of information technologies like for example the ERP systems in mining find it hard to keep up with many real-time aspects of mining (KPMG, 2016). To make informed business decisions, it is important to have the right information, at the right time, to the right person. Right decisions can be made if the right information is gathered, analyzed, reported, and applied to the business.

Mining companies usually have a large amount of data from drills, trucks, processing plants, etc., but in many cases, less than 1% of these data are used for any kind of decision-making (Whyte et al., 2015). Investments in systems and tools help to build a foundation for better decision-making. The data provided by the systems are helpful in being used with decision-making algorithms to enhance better decision-making. In mining operations, the use of information technologies aids in decision-making across multiple job functions. Some of the important decision-making factors carried out by mining personnel efficiently with the help of specialized technologies include (McRoberts, 2016):

(a) tracking equipment operators for overall equipment efficiency and mean time between failures,
(b) reduction in unexpected downtime occurrences done by maintenance technicians to efficiently monitor the asset health regularly and predictive maintenance,
(c) reviewing the ore grade and quality of the product by quality managers,
(d) viewing of crucial factors like real-time cost of production by site managers, and
(e) comparisons between real-time operations and commodity prices for adjustments to be done accordingly, done by top executives.

Recent technologies that are server-based solutions such as dispatch systems and production management systems can support decision-making in operations across mine sites and can increase productivity (Thompson, 2015). Many mining companies do not integrate their existing systems with modern systems and are not able to benefit from modern technologies. To reduce the system integration gap that can improve decision-making, an end-to-end approach must be taken in mining organizations (EY, 2017). Although the mining industry has embraced the new information technologies to bridge the digital disconnect between the systems, the rate of development is slow in comparison to the opportunities available.

5.2.6 *Interrelationship Between Information Technologies and Productivity*

Technology is becoming increasingly important for mining companies to adapt and keep up with the competitive advantage factor (PwC, 2016). Productivity is the number one operational opportunity in mining and being digital has potential benefits for mining companies. Ernst & Young, a professional consulting service provider, conducted a survey of global mining leaders (EY, 2017). The survey concluded that new thinking along with a better focus on generating value will enhance performance and information technology is the key to harness productivity opportunities in the mining sector. Information technology should be used as a strategic tool for enabling better-operating models and practices and it requires mining organizations to make the right technology investments (Thompson, 2015).

Technology plays a positive role in enhancing productivity by helping in resolving issues that used to take hours to be done very swiftly (Mining Global, 2015). There are other productivity gains due to automation like a decrease in maintenance cost and the distance between the two machines can be reduced thus increasing operational efficiency. Autonomous technologies like driverless vehicles, remove the errors which take place due to human performance and increases the efficiency of equipment utilization by reducing machine idle time and consistent speed in driving which in turn helps in improving efficiency in fuel consumption (McNab & Garcia-Vasquez, 2011).

Mining companies across the world are pioneering toward implementing best business practices by using the latest-proven information technology systems. A revolutionary new technology known as the Blockchain technology, which was previously only used by other manufacturing companies, is now being used in mining companies. The Blockchain technology is already used by diamond mining companies as a quality assurance measure to verify quality, ethical extraction, and authenticity (PwC, 2017).

To achieve operational excellence, it is not enough just to cut costs and improve grade recoveries, mining companies need to harness the full potential of innovated technologies like a simulation of scenarios, driverless equipment, and big data analytics suites. Operational efficiency can be achieved by automation and integration of systems through the supply chain (Business First Magazine, 2017). Mining deals with many remote locations and new technologies and analytic solutions help to increase operational performance by getting information at the right time, thus helping better transparency of information in the mining organizations in real-time (Business First Magazine, 2017). Real-time data and better data analytics help in processing decisions faster, which in turn helps in productivity factors like equipment usage, understanding machine movements to maximize efficiency.

One important benefit of real-time data is knowing the location and state of each piece of equipment in mining operations and whether it is operated according to plan. New advanced technologies, like the operational effectiveness analytics solution developed by IBM's Haifa Research Lab, address operational excellence in the mining industry by identifying defects and minimizing process variabilities and can be customized for any type of open-pit mining operation.

Remote mining operations, if well connected and integrated, can also achieve operational excellence and one such example is the IROC in Australia, of the company BHP Bilton. The remote-access technology of the company offers experts the opportunity to work from a single location and provide support to the companies' branches worldwide, including remote monitoring of equipment and alerting on-site personnel about the important issue, or sometimes virtually logging in to address the issue. In many mining companies, the data already exist, but it has not been changed to information, which can be useful, and data need to be collected, analyzed, and shared to change it into important information.

Due to the growth and large investments in the mining sector, the dependency on mining equipment and machinery is huge (E&MJ, 2012b). Equipment and machinery require maintenance and this, in turn, requires the manufacturers to come up with solutions that can integrate with the mining enterprise systems (E&MJ, 2012b). Equipment utilization is maximized due to less downtime related to breaks and shift changes and the costs related to physical damage caused by human error, such as collisions and wear-and-tear of tires, are also removed.

Technology can now help in analyzing the existing data and provide a set of data visualizing reports to show inefficiencies taking place in mining operations, which can be used by mining personnel to understand and rectify the inefficiencies. One of the large companies in information technologies, IBM, is using its full potential in analytical capabilities behind the mining companies to help increase production and make changes in the process to run it efficiently. A breakthrough in productivity performance requires a change in the way of thinking about how mining works, which are where technological innovations play a pivotal role in changing key aspects of mining.

A report by Ernst & Young stated that the global mining sector faces many challenges or risks in the form of resource nationalism, worker shortages, and infrastructure access (Procom, 2012). However, technology should be aligned to the business needs, which is not usually the case in most mining companies and thus the mining industry has a bad reputation for not delivering business value. Today, some companies run multiple instances of ERP systems across multiple departments, some types are single-user desktop software supporting technical tasks. It is complex and expensive to acquire, implement, and maintain technology, and mining companies do realize this fact.

Investments in technology add up to the operating cost, therefore, the mining companies must drive value from the investments. The companies can do cost cutting by investing more in innovation in terms of automation in drilling systems, data analytics, and mobile technologies, and this, in turn, reduce the intensity in people, capital, and energy. The potential to reduce operating costs and improving operating discipline are some advantages of automation. In open-pit mining, costs are mainly affected by the number and capacity of equipment, and the selection of equipment is a huge decision that will affect the economic viability of mining operations greatly.

Choosing a cost model for equipment selection is another challenge since many types of models exist, for example, traditionally, the major costs of looking after the equipment during its useful life were not considered and the modern cost model uses expert systems for decision-making in open-pit mine equipment selection or the technique of net present value analysis applied to equipment selection. Some cost models for equipment selection were done while other cost models are based on computer software, like the EQS that uses fuzzy logic, which is based on maximizing production and minimizing the unit stripping cost. The mining sector uses a wide variety of information technologies to its advantage. It is also revealed that huge investments are made by mining companies on advanced information technologies.

5.3 Theoretical Contribution for Operational Excellence in LSOPM Operations

The theoretical contribution for operational excellence in LSOPM operations was created after combining the outcomes of previous studies and the current research work. Three broad frameworks were created initially to better understand the components leading to the final conceptual model. The three frameworks are in the form of:

- The Prescriptive Framework,
- The Descriptive Framework, and
- The Normative Framework.

5.4 The Prescriptive, the Normative, and the Descriptive Framework for LSOPM Operations

In terms of the ***Prescriptive Framework (PF)***, which is based on previous studies and the current study, a combination of both the outcomes resulted in four important factors forming the most important requirements for an LSOPM operation. The factors and their relevance are described below:

- *Understanding the mine deposits*
  a. Planning for the various phases of mining according to the deposit for cost efficiency purposes.
  b. Modeling the deposit into block models to mine them according to the type of ore block and the different types of grades during recovery for improving productivity.

- Better decision-making processes
    a. The relevance of decision-making in an organization in terms of accountability and responsibility.
    b. The distinction between strategic and operational decisions according to levels of management.
    c. The relevance of decision-making in equipment selection, procurement, and use.
    d. The relevance of decision-making in processing and grade control.
    e. Better decision-making processes required.
    f. Better decision-making processes can result in improving organizational efficiencies, strategic decision-making, the swiftness of decision-making, and productivity.
- Improving Organizational Efficiencies
    a. Cost optimization is a key concept. It can be related to cost reduction but is not the same as cost reduction.
    b. Productivity improvement in the form of efficiency in equipment, manpower, performance, production, and time are all factors that were repeated by many respondents. Process efficiency has been mentioned by many respondents as one of the important organizational efficiencies. The emphasis was more on equipment, manpower, and performance efficiencies. Production efficiency is the least in the order of mention by respondents.
- Use of Information Technologies
    a. Information availability and automation have been strongly recommended for every phase of mining in an LSOPM operation.
    b. Real-time information availability and monitoring are also included as a key requirement by the respondents.
    c. Technological innovation and breakthrough in mining include:

- Remotely operated driverless trucks
- Expert systems
- Mine planning software
- Geological software
- Fleet management systems
- Software technologies for drilling and blasting
- Collision awareness system
- Survey software

The organizational efficiencies that form one of the important outcomes of the study and that are related to productivity improvements are further detailed below:

- *Equipment Efficiency:* The keyword for this efficiency is equipment. The efficiency in equipment in the LSOPM operations can be determined by the capacity of the equipment, utilization based on the availability of the equipment and the maintenance practices followed for the upkeeping of the equipment. A particularly important factor is also the purchase of the equipment from the OEM, a decision that depends on the trade-off between cost and type of requirement of a mining operation.

- *Manpower Efficiency*: The keyword for this efficiency is manpower or human resources. The efficiency of manpower or labor productivity can be determined in terms of output per capita, output per person employed, and output per employee hour. The research, however, shows that the efficiency in manpower depends on many factors such as skills, knowledge, experience, qualification, which determine productive hours spent in the operation. The efficiency of manpower can be improved by professional development and training.
- *Performance Efficiency*: The keyword for this efficiency is performance. Performance efficiency deals with implementing performance measurement systems to enhance the performance of all organizational resources by monitoring and controlling it. The systems for measuring performance include setting up benchmarks and targets by using Key Performance Indicators (KPI's) in all areas of mining.
- *Process Efficiency:* The keyword for this efficiency is the process or processes. Process efficiency is related to how well organized the processes are in an LSOPM operation. The PIO methodology for process efficiency is a very useful methodology to improve productivity by increasing efficiency in processes. The responses relating to process efficiency can be summarized into the following: sustainability of processes, realizing bottlenecks in processes, simplifying processes, optimizing processes, disruptions in processes impact unit costs, and process optimization for improving equipment utilization.
- *Time Efficiency:* The keyword for this efficiency is time. The efficiency in time does not relate to only the factor of producing more in less amount of time. The research implies that saving time is an important factor and needs to be considered in terms of decision-making in organizations, equipment efficiency, or performance efficiency.
- *Production Efficiency:* The keyword for this efficiency is output. Efficiency in production is not the same as productivity. While productivity includes a broad spectrum of using resources efficiently, production efficiency deals with improvements in production or an increase in production using various techniques. An example of this can be taken from a few of the responses in the research in terms of "decreasing production increases costs" or "reducing costs does not ensure an increase in production." Therefore, production efficiency can be concluded as one of the subdefinitions of productivity and related to cost optimization.

The ***Normative Framework (NF)*** is created taking the most prominent outcomes combining the previous studies and the current research and it is given in Fig. 5.2. It

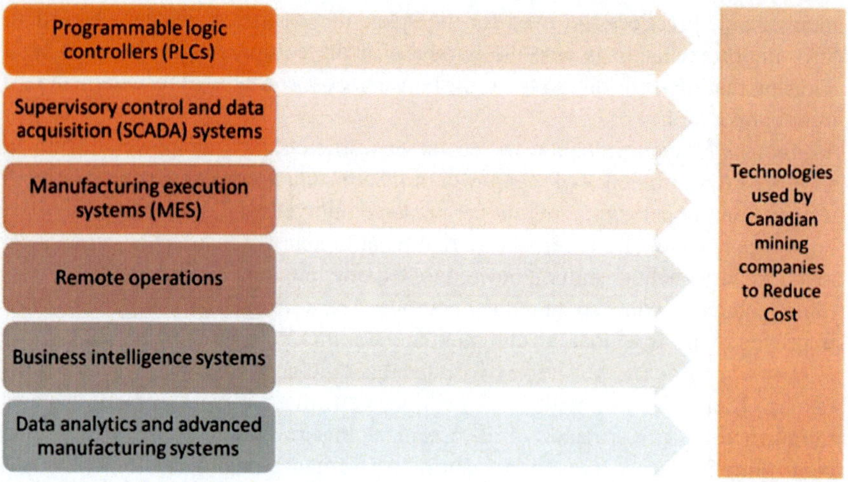

Fig. 5.2 The normative framework

contains three components in terms of information technology, productivity improvements, and cost optimization. The three components form the core components of the LSOPM requirements and are further detailed below:

- *Information Technology*

Research has shown the significant importance of information technologies in LSOPM operations. Expert systems that help in operational and strategic decision-making; or driverless technologies that help in increasing operational efficiencies; or fleet management systems that help in day-to-day operations in the open-pit are all examples of systems that assist in both operational and strategic decision-making.

- *Productivity Improvements*

The improvements in productivity are related to organizational efficiencies. The prominent factors to improve productivity are manpower efficiency, equipment efficiency, time efficiency, and digitization or automation.

- *Cost Optimization*

The price of metals and minerals is determined by the market and this is one of the most important reasons why when decision-making is slower, it affects productivity and the economic viability of an LSOPM project. Improvement in production, efficiency in time, efficiency in equipment utilization, efficiency in the use of manpower resources, cost optimization, increased performance, and profitability all lead to an improvement in productivity as shown in Fig. 4.4. The inner circle includes the factors that are most relevant toward the effects of swift decision-making on productivity and the outer circle depicts the factors, which have a lesser effect on productivity in comparison to the inner circle factors. The cost needs to be managed. A remarkably interesting finding shown in Chap. 2 on the relationship of cost with productivity shows that increasing productivity can decrease cost, however, it is also important to note that cost reduction is not the same as cost optimization. The reduction in cost must add value in terms of productivity.

The ***Descriptive Framework (DF)*** contains the broad outcomes less mentioned in literature and the research but particularly important factors leading to operational excellence in LSOPM operations. The four important factors are:

- *Better Understanding of Mine deposits*

  a. The mine deposits need to be understood in the exploration stage very well in terms of economic viability, the life of mine, and the type of ore available.
  b. As operations continue, the mining model needs to be developed in a dynamic way to understand which block to mine and if the blocks are ore blocks or waste blocks and decide on blocks in terms of cut-off grade.

- Risk minimization

  The mining companies need to keep in check the risk involved in terms of:

a. capital expenditures for equipment procurement.
b. resource nationalism, worker shortages, and infrastructure access, and
c. risks to be considered while choosing an alternative in decision-making.

- Effective communications

  a. Communication systems should be integrated into automation.
  b. Short mine communications are important to be effective for day-to-day operations to run successfully.

- Safety

An especially important factor that is related to *Safety*, otherwise also known as Health, Safety, and Environment in the mining industry is not included as a factor as it requires to be studied further as a separate topic. It is important for the sustainability of the mining operations and the environment around it affecting all the stakeholders equally.

5.4.1 The PMLR Model

The prescriptive, normative, and the descriptive frameworks mentioned above in Sect. 5.2.1 leads to the creation of the new conceptual model. The model created by integrating the three frameworks is the Prototypical Model for LSOPM Requirements (PMLR), which is illustrated in Fig. 5.3. The PMLR model consists of four rings.

The four rings of the PMLR model are prioritized based on the most important requirements for an LSOPM operation from the innermost circle to the outermost

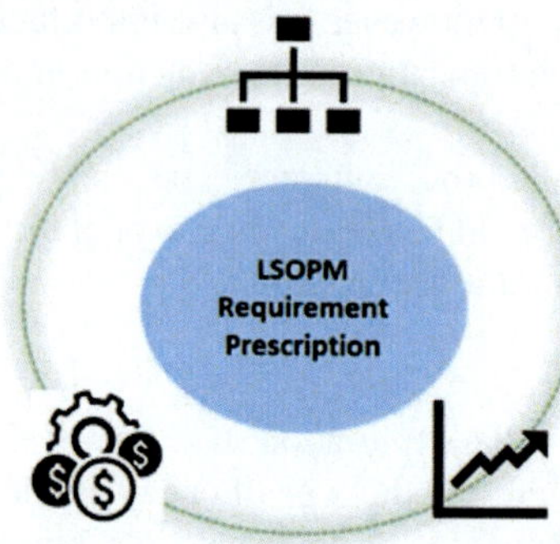

DOíS

- Integrate all information technologies implemented into a single connected system.
- Optimize costs from the time the deposit is discovered and found to be economically viable.
- Optimize costs throughout the entire value chain to increase overall productivity.

DONTíS

- Invest in information technologies unless the exact change in costs and productivity are measured.
- Reduce cost without realizing improvement in productivity.
- Increase cost without realizing the value that the increase will bring.

LEGEND

1. = Information Technologies
2. = Cost Optimization
3. = Productivity Improvement

Fig. 5.3 The prototypical model for LSOPM requirements (PMLR)

rings. Each ring is composed of requirements that have been revealed based on the study.

1. Ring one: Ring one is the innermost ring and depicts the core requirement of LSOPM operations as per the NF model in Fig. 5.2. Information technologies, cost optimization, and productivity improvements form the three pillars that enhance the overall efficiency of the operations. The inner circle components are not stand alone and work as tools for the overall resources to be used in a way that ensures profitability and revenue from the LSOPM operations.
2. Ring two: Ring two of the PMLR model extends to decision-making, equipment efficiency, and manpower efficiency. The components in ring two further concentrate on the resource and technical efficiency of the LSOPM operations and focus more on the two important resources manpower and equipment to which decision-making is related. This is in line with the decision-making and organizational efficiency concepts mentioned in the PF model in Sect. 5.2.1. The components of ring one need to be incorporated into the components in ring two to be made useful.
3. Ring Three: Ring three of the model focuses on three important components in any organization, process efficiency, time efficiency, and performance efficiency. This is a continuation of the organizational efficiencies mentioned in the PF model in Sect. 5.2.1. Time saving and process efficiency are both related to cost saving and both together ensure performance efficiency. The components in ring

one and ring two need to be incorporated for the components in ring three to be made useful.

4. Ring Four: Ring four focuses on production efficiency, a better understanding of ore deposits, effective communication, and minimizing risk. The four components are interrelated and are effective if the inner core components are effective. Production efficiency was the least mentioned by the respondents as already mentioned in Sect. 5.2.1. The DF model in Sect. 5.2.1 refers to the other three components of the prototypical model. Understanding the ore deposits is one of the components that is important but was mentioned less in all studies. It is important to emphasize that the requirement of understanding the deposits in terms of demand in metals, type of ore deposits, and modeling of ore deposits play an important role in the sustainability of mining operations. The two components of minimizing risk and effective communication are an outcome of the primary research and are important considering the risk in investments and the scale of economies of the industry. The components in ring one, ring two, and ring three need to be incorporated into the components of ring four to be made useful.

5.4.2 Outcomes to Be Replicated for Other Organizations

The LSOPM requirements are requirements that can be morphed according to the conditions of other fields. Every field and organization has the use of almost all the requirements mentioned in the LSOPM requirements. Information technology is considered a revolution from the time it came into existence and its importance has already been established in Chap. 3. Other fields have more developed technologies than LSOPM organizations. However, to what level is its contribution to the business needs to be further analyzed, as per the type of business in the study.

Productivity improvements and cost optimization relate to operational excellence and is a leading enabler for all organizations. The relationship between cost and productivity clearly stated in Chap. 2 is a general theory that can be applied to all types of organizations. The efficiencies in the form of equipment use, manpower use, cost, time, process, and overall performance can also be considered for all types of organizations. Decision-making is valuable, and a better understanding of the decision-making process is beneficial for any organization to sustain the business. An increase in production is something that needs to be understood based on all other factors.

The only factor that may not be relatable to other organizations is the factor of understanding the mine deposits better. The factor focuses on the mining scenario and is relatable to all kinds of mining operations including fossil fuel mining and underground mining but does not relate to other kinds of businesses. The use of current research in other fields is depicted in Fig. 5.4.

*LSOPM Requirement "Better understanding of Deposits"—not applicable to any other organization.

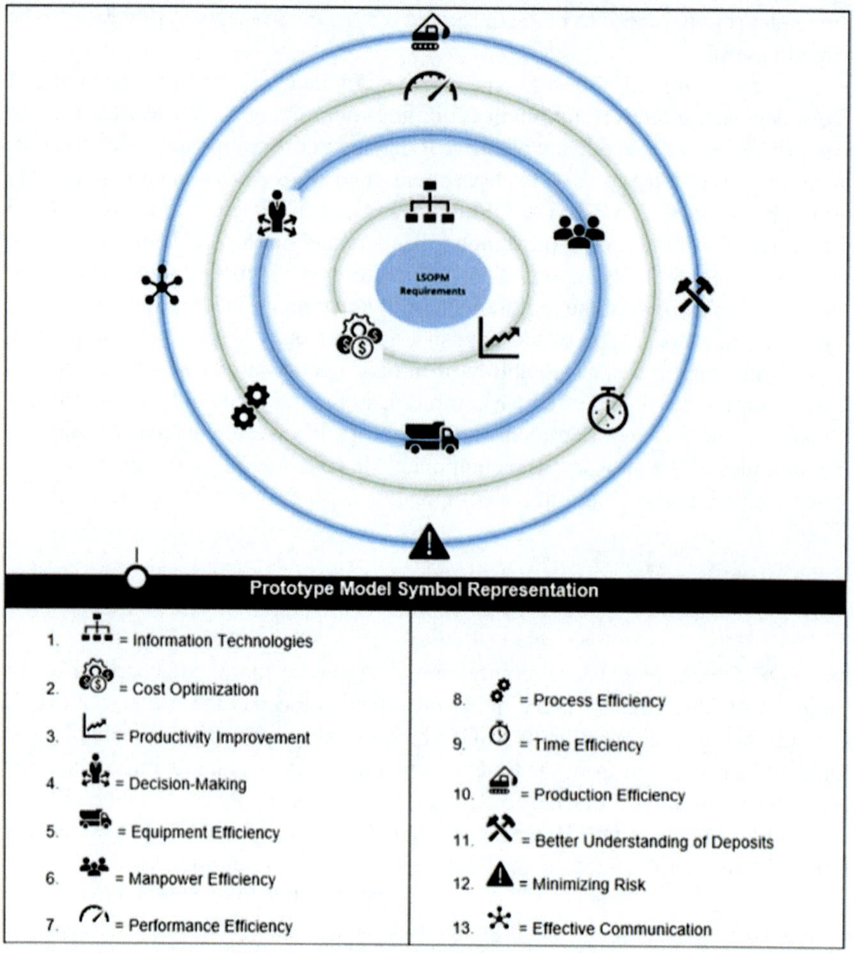

Fig. 5.4 The use of LSOPM requirements in other areas of business

It is expected that the study will aid the managers and employees working in the LSOPM operations in the top-level, middle-level, and the first-line management levels to:

- Understand the core requirements of an LSOPM operation according to the PMLR model and plan how to achieve it,
- Successfully implement tools and techniques according to the understanding of the above requirements and planning,
- Measure the implemented techniques with the planned ones to see if the requirements have been followed accordingly, and
- Control the operations as per the requirement of the model.

5.5 On the Scope and Relevance of This Research

With higher demand for metals and minerals in our day-to-day life, it is required that the mining field is regarded as an important field where not just engineering and technology, but a combination of engineering, technology, and management study can bring in a lot of desired changes. There is a huge gap between the industry and the academic field of mining. The two important factors, risk management and health, safety, and environment have not been included in detail in the book because both these factors require a detailed study of its own.

Risk management in mining

Risk management includes and looks beyond factors like talent management, labor relations, small-scale mining that affects production, community relations, getting additional permits, etc. (Mining Magazine, 2016). Risks can be production related or financially related and it should be managed by understanding the risks that each department can have so that the risks do not materialize. Uncertainty and financial risk are part of a mining project from the time the initial exploration and feasibility studies start until the closure of the mines. Every stage has a time when a decision to stop or continue further must be made based on the information available at that time (Botin et al., 2011).

An interesting methodological framework is developed by collaborative research effort between the Universidad Politecnica de Madrid (School of Mines) and Pontificia Universidad Católica de Chile (Mining Center) known as the Mining project risk management (MPRM) (Botin et al., 2011). This framework concentrates on identifying and managing decision risks along the project value chain. According to this framework, risk can be classified based on economic impact and probability of occurrences into four classes (Botin et al., 2011).

(i) Class A—high-probability/high-impact risks
(ii) Class B—low-impact/high-probability risks
(iii) Class C—low-probability/high-impact risks
(iv) Class D—low-impact/low-probability risks

Health, safety, and environment

The health of employees is an especially important factor to enhance employee efficiency. For example, if a crew has 10 employees out of whom 3 are sick, this could mean 3 trucks are parked, which affects production and ultimately productivity.

Safety relates to work conditions, which do not affect the health of the employees. Any incident in the work area will directly influence the health of the employees, which affects production and, in the end, affects productivity. Safety should be of number one priority when employees are concerned.

The environment is of grave importance in the mining scenario. For example, the release of toxic substances could mean damage to biodiversity, tarnishing the reputation of the organization, and even sometimes causing the cancellation of license to mine (Ontario, 2020). Mining organization needs to be environment friendly and work toward achieving accolades for proper environment care.

The safety challenges to the mining industry are unique and can be faced by identifying and minimizing potential hazards relating to the environment and equipment (Saguaro et al., 2020). The potential of loss in life and environmental damage is much greater in the mining industry compared with any other industry. As the operations get bigger and bigger the challenges faced getting bigger too which requires corporates to come up with specialized safety standards and practices and equipment, which are embedded with alert and warning systems to minimize accidents in the mining pit.

A safe mine environment can be created through the combination of technologically advanced equipment and rules and regulations, which enhance safety through precautions (Saguaro et al., 2020). Mineral extraction and transportation in the open-pit through various automated technologies and conveyor belts and the use of GPS and collision awareness systems have brought down the risk of employee injuries in many operations. The rules and regulations in many countries in the world are not very distinct for mining, which puts more pressure on mining organizations to follow international standards and have stringent safety policies followed strictly by all employees in the organization.

In terms of safety, another area of challenge is in the case of explosives used for blasting (Ontario, 2020). The storage, transportation, and use of explosives must be planned carefully and followed accordingly. The storage of explosives requires special conditions to protect it from elements and conditions that could degrade its quality or reliability causing a hazardous situation at the time of its use. It should be available only to authorize trained personnel who would use it only for the intended purpose (Ontario, 2020). There should be strict safety practices in using explosives such as (Ontario, 2020):

a. timely proper maintenance of equipment using explosives and place of storage of the explosives.
b. records of quantities of explosives purchased and quantities used from the inventory must be managed accurately along with checking the condition of the explosives if it has not been used for a certain period after purchase and store.
c. in case of damage, specific procedures must be followed to dispose of the explosives without causing any harm to people and the environment.
d. handling of explosives during transportation from warehouse to pit must be done through the standard procedure and care should be taken that it is never left unattended.

In terms of maintaining safety from the use of mining equipment, many times the reason is that of collision incidents in the pit. Most of it can be controlled through the development of traffic management program in the mining area using policies in maintaining traffic and putting measures in places like the use of reflective clothing, well-illuminated vehicles and roads, proper monitoring of the area through technologies, and proper training to employees concerning following the traffic rules and regulations in the mine area (Ontario, 2020).

Mining organizations are sensitive to the social and environmental components of the business. The global community looks at mining organizations to impact the environment positively and remove the previous notion of the people that mining destroys the environment and affects the place and the people in the mining area negatively. They need to have the social license to bring in positive change to the communities that live near the mine area and deal with sensitivity on issues relating to illegal mining done by the local people due to this being something they have been engaged in for generations but since now the land is licensed to an organization this is now considered illegal. Management in such mining organizations should work with the communities to develop them and help them acquire skills and to assist in employment in the organization or elsewhere. This can be done by hiring people from the local community, buying local supplies from them so that they benefit financially from the project, etc.

For instance, if there is water in the pit that needs to be discharged to the environment, it needs to be checked that the quality of the water is good and will not cause any harm to the environment. This is crucial for health, safety, and the environment. Information technology can help in accessing data to get the result of sampling of the water to know if the quality of the water is good enough for it to be discharged to the environment which is a crucial decision a manager has to take. There are information systems that help in deciding which area is suitable for the water discharge and which area is not suitable and making such decisions quicker helps in increasing productivity.

Trucks in the mine area are not allowed to be driven if there is too much dust on the roads, this requires proper dust controlling measures for which there are software systems that help in letting the crew know when the dust amount requires controlling so the managers can decide accordingly. Health and safety are closely related to production. When an incident occurs and even though it is not a recordable incident—meaning someone did not get hurt—still there is some cost to equipment repair, but safety is ensuring that there is no damage of any kind not even to the equipment because such damage can be taken as a warning signal for a more serious incident in the future.

5.5.1 *Future Research Directions*

It is expected that the research encourages more studies dealing with the above recommendations and conclusions to develop new ideas and models which can be successfully tested in the mining field. The focus of the research is on understanding the requirements for large-scale open-pit mining operations. The study can be continued in the future to uncover concepts on the following:

1. Going beyond management level: The research uses a sample from various levels of management but does not include all types of ground-level staff and equipment operators. A detailed study on how the ground operators find the use of

information technologies aiding in reaching day-to-day operational tasks can be undertaken. Such a study could enhance the ground-level realities of high-end technologies in large-scale open-pit mining operations.

2. Comparison between management level and ground level staff on the use of information technologies in LSOP operations: A detailed study on the relevance of information technologies in LSOPM by comparing the views of the management-level employees to the ground-level staff can be undertaken. Such a study could give an overview of the similarities and differences of opinion between the two different groups of samples.
3. The relevance of information technologies in upkeeping environmental and safety standards in LSOPM: The research does not delve into the safety standards in detail for the LSOPM operations. The environmental impact of mining and the adherence to various norms and regulations to have a minimum negative effect on the environment are issues that need to be addressed on a stand-alone basis. The current study can be further continued keeping in mind how much of the EIA (Environmental Impact Assessment) is done by mining companies and its importance in LSOPM operations along with how information technologies can assist in adhering to the safety standards.
4. LSOPM operations for fossil fuel: Fossil fuels like coal, gasoline, petrol, diesel, etc., are also mined using the open-pit mining method but differs a lot from the mining of metal and minerals. There was thus a requirement of separating it from the current study. A study on the requirements for LSOPM operations of fossil fuel can be continued as future research to understand how it differs from the current research.
5. Underground Operations: The research addresses the requirements of open-pit mining requirements. A study on the underground mining requirements can be further continued as there are similarities between the two but there are differences too, which is the topic of underground mining operations is not included in the research. The differences are evident in the type of mining methods, the risks involved, and the resource skills required for underground mining operations.
6. Testing of the PMLR model: The PMLR model developed in the research is a prototype and has not yet been tested on a real-LSOPM operation. The model may have its discrepancies that can be realized only after a similar study is done using another set of samples from various other countries or nationalities with organizations from various other locations. The model cannot be guaranteed to be used in all the LSOPM operations in the world because every operation has its challenges and factors, which allow it to continue as a business.

5.6 Conclusion

After a thorough analysis of the results, it has been revealed that for an LSOPM operation, there are many requirements but information technologies, cost optimization, and productivity improvements form the core requirements as were determined through the NF model developed in Fig. 5.3. However, the results also show that it is indicative that other important requirements need to be focused on too to run the LSOPM operations successfully. Many studies reveal that the research conducted is more focused on the mining methods, rather than focusing on open-pit mining scenario only when it comes to its core requirements in techniques. The key take-away from each chapter is given in Table 5.1 in the appendix section at the end of the chapter.

In conclusion, the current research has made the following four observations about LSOPM operations:

- *Use of Information Technologies are critical*

The type of current information technologies available for LSOPM operations given in Chap. 3 assists in understanding the contribution that information technologies to many areas of the LSOPM operations. Every other requirement mentioned for the LSOPM operations in the research is directly enabled to be made better with the use of information technologies. It is evident from the research that the availability of information and the efficient use of information technologies play an overly critical role. The research also reveals that the inefficient use of information technologies can affect the organization negatively. The general technologies like email, Skype, Microsoft Excel, and SharePoint and database systems and ERP systems used in general by other organizations are mentioned along with the specialized software for specific areas of LSOPM operations, in the current study. All the mentioned systems, when used efficiently, can enhance the productivity of LSOPM operations.

- *The relationship between costs and productivity*

An interesting relationship between productivity and cost has been observed in the current study. A decrease in productivity causes an increase in cost and an increase in productivity shows that cost has been decreased. This theory works most of the time even though exceptions can occur. Cost efficiency deals with various costs and the requirement to understand which are the costs that can be considered for trade-offs to increase the overall cost efficiencies. When the same resources or fewer resources are used to increase production, the cost is reduced, and productivity is increased. However, when more resources increase production, the ratio of cost of the resources and the increase in production should be measured to realize if productivity has decreased or increased. If the increase is not more in production in comparison to the cost of the resources, then productivity decreases, and this proves the theory right.

- *Decision-making processes need to be efficient*

The decision-making process is a key contributor to any organization and needs to be efficient. If the decision-making process is faster, the decision-making can also be faster. The efficiency in decision-making processes accelerates the efficiencies in manpower usage, equipment usage, time, and cost. It helps in strategic decision-making in the upper level management for the long-term sustainability of the business. Efficiency in decision-making processes can assist in improving productivity and increasing the profitability of LSOPM operations.

- *The importance of key organizational efficiencies*

The research has revealed six key efficiencies that should be the focus of LSOPM operations. An equipment-centric business like LSOPM operations requires strategies and planning in equipment selection, equipment maintenance, equipment availability, and equipment utilization. Manpower efficiency deals with the deployment of the right people at the right positions, skill development through training, and employee attitude and behavior in the organization. Time efficiency is related to a decrease in idle times or waits times of equipment, a faster rate of production, faster processes, and faster decision-making. Process efficiency deals with improvements required to make processes faster through faster decisions and the right technologies to assist in it. Performance efficiency deals with benchmarks, targets, goals, and budgets. An important terminology used in this regard is for operations to perform efficiently to achieve the Key Performance Indicators or KPI's. Apart from that, the final factor is volumes produced efficiently also referred to as production efficiency.

5.7 Summary

For an LSOPM operation, there are many requirements but information technologies, cost optimization and productivity improvements form the core requirements. However, the results also show that it is indicative that other important requirements need to be focused on too to run the LSOPM operations successfully.

The present study has unveiled the following findings:

- Understanding how the concepts of production, productivity, and cost optimization are defined within the profession.
- In-depth understanding of what constitutes as factors for improving productivity and drivers of cost optimization.
- Examining the effect of productivity change on cost optimization and understanding the role of equipment utilization in productivity improvements.
- Understanding the impact of information technologies on productivity and decision-making.
- Creation of a conceptual model through a combination of prescriptive, normative, and descriptive frameworks, which shows the importance of the concepts and its relationship to LSOPM requirements.

Mining is a field that is important due to the importance of metal and minerals in our day-to-day life. However, traditional mining methods differ from the conventional and technical methods that have been embraced for a while now. Also, decision-making along with the efficiency of equipment, manpower, time, cost, process, and overall performance plays a vital role in making the LSOPM operations successful. The requirements mentioned in this book in chapter 5 provide a clear picture of how an LSOPM operation needs to be managed keeping of primary importance the long-term sustainability factor.

Appendix: Additional Information on Key Takeaways from Chapters

See Table 5.1.

Table 5.1 Key takeaways from each chapter

Chapter #	Key takeaway	Description
1	Understanding the mine deposits	The first chapter relates to all aspects of LSOPM like the following: 1. The relevance of mining, planning for the various phases of mining according to the deposit, 2. mining processes, 3. modeling the deposit into block models to mine them according to the type of ore block and the different types of grades during recovery
2	Value added by information technologies	Information availability and automation to increase efficiency are the key requirements of an organization. This chapter focuses on technological innovation and breakthrough in mining, which includes: 1. Remotely operated driverless trucks 2. Expert systems 3. Mine planning software 4. Geological software 5. Fleet management systems 6. Software technologies for drilling and blasting
3	The relevance of cost optimization, productivity, equipment, and organizational efficiencies	The third chapter focuses on the relevance of many factors related to organizational efficiencies in the form of: 1. Cost efficiency/cost optimization 2. Productivity in the form of equipment efficiency, manpower efficiency, performance efficiency, production improvement, process efficiency, and time efficiency

(continued)

Table 5.1 (continued)

Chapter #	Key takeaway	Description
4	Better decision-making processes	The fourth chapter focuses on: 1. decision-making relevance in equipment selection and grade control 2. the distinction between strategic and operational decisions according to levels of management
5	Requirements of LSOPM operations	A summary of all the conclusive findings from each chapter is provided in this chapter. The following are the main focus of the chapter: 1. The Prototypical Model of LSOPM Requirements is introduced showing emphasizing the relevance of 13 factors for LSOPM operations 2. Future research directions

References

Australian Mining. (2016, August 17). *The top ten trends for mining in 2016: Australian mining.* Retrieved from Australian Mining Web site: www.australianmining.com.au.

Barr, R., & Cook, E. R. (2009, November). Mining for operational excellence. *Mining Engineering,* 17–22.

Bazzazi, A. G., Osanloo, M., & Behrooz, K. (2009). Optimal open pit mining equipment selection using fuzzy mutiple attribute decision making approach. *Archives of Mining Sciences, 54*(2), 301–320. Retrieved from https://mining.archives.pl/.

Botin, A. J., Guzman, R. R., & Smith, L. M. (2011). A Methodological model to assist in the optimization and risk management of mining Investment Decisions. *Scielo.*

Business First Magazine. (2017). *Business operational excellence: Business First Magazine.* Retrieved from Business First Magazine Web site: https://www.businessfirstmagazine.com.au.

Deloitte. (2015). *Tracking the trends 2016: The top ten issues mining companies will face in the coming year.* Deloitte.

E&MJ. (2009). New tools put data to work. *Engineering & Mining Journal*, 68–71.

E&MJ. (2012, March). Chilean service companies become world class. *Engineering & Mining Journal*, 68–80.

EY. (2017). *The digital mine: EY*. Retrieved from EY Website: www.ey.com.

Hanson, D. (2016, June 1). *Problem solving: A look at analyzing mining data to improve performance.* Retrieved from Canadian Mining Journal Web site: www.canadianminingjournal.com.

ICMM. (2012a). *The role of mining in national economies.* International Council on Mining & Metals. Retrieved from https://www.icmm.com/document/4440.

KPMG. (2016). *Building confidence: A review of Dubai's residential real estate market.* KPMG. Retrieved from https://assets.kpmg/content/dam/kpmg/pdf/2016/04/building-confidence-uae.pdf.

Lamghari, A., Dimitrakopoulos, R., & Ferland, A. J. (2013, July 3). A variable neighbourhood descent algorithm for the open-pit mine production scheduling problem with metal uncertainty. *Journal of the Operational Research Society*, 1305–1314.

McNab, K., & Garcia-Vasquez, M. (2011). *Autonomous and remote operation technologies in Australian mining.* Brisbane: The University of Queensland.

McRoberts, P. (2016, September). World mining equipment: Operating strategies-counting on the connected mine. *Engineering & Mining Journal*, 158–164.

Mining Global. (2015, August 12). *GoldCorp—Operational excellence done right: Mining global*. Retrieved from Mining Global Web site: www.miningglobal.com.

Mining Magazine. (2016, July 14). *What's impacting productivity in your operation?* Retrieved from Mining Magazine Web site: https://www.miningmagazine.com/innovation/news/1263123/%E2%80%99-impacting-productivity-operation.

Ontario. (2020, July 02). *Key hazards in mining health and safety*. Retrieved from Ontario Government Web site: https://www.ontario.ca/document/mining-sector-plan-2017-2018/key-hazards-mining-health-and-safety.

Procom. (2012, December 7). *Information technology helps spur innovation in Canadian mining industry*. Retrieved from Procom Web site: https://www.procom.ca.

PwC. (2016). *Mine 2016 slower, lower, weaker... but not defeated: PricewaterhouseCoopers*. Retrieved from PricewaterhouseCoopers Web site: www.pwc.com.

PwC. (2017). *We need to talk about the future of mining: PricewaterhouseCoopers*. Retrieved from PricewaterhouseCoopers Web site: www.pwc.com.

Saguaro Conveyor Equipment. (2020, July 02). *The importance of proper mine safety, and how it's done*. Retrieved from Saguaro Conveyor Equipment Web site: https://saguaroconveyor.com/the-importance-of-proper-mine-safety-and-how-its-done/.

Thompson, I. (2015, December 1). *Harnessing information technology to drive productivity gains: Canadian Mining Journal*. Retrieved from Canadian Mining Journal Web site: www.canadianminingjournal.com.

Whyte, D. H., Geraghty, R., Pujol, F., & Sellschop, R. (2015, November). *How digital innovation can improve mining productivity: McKinsey & Company*. Retrieved from McKinsey & Company Web site: www.mckinsey.com.